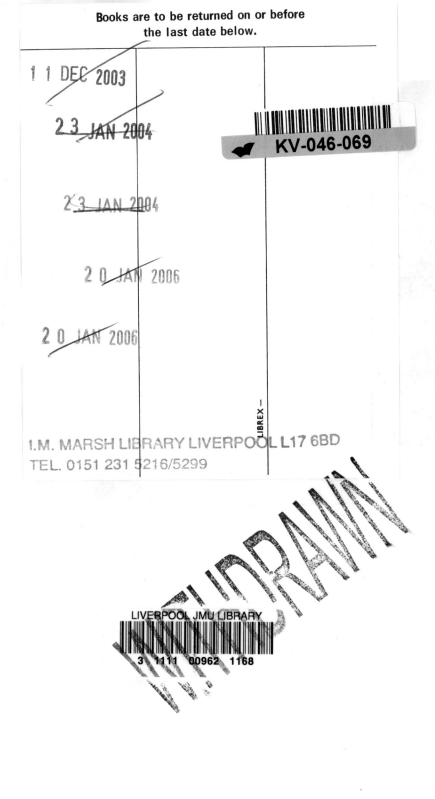

# Lessons for the Future

If one of the main purposes of education is to prepare young people for the future, then where in education are they given the opportunity to explore the future? In this thought-provoking book, David Hicks shows the future to be a neglected dimension in education and argues elegantly and persuasively for a more futures-orientated curriculum.

Drawing on some of the latest research in futures studies, *Lessons for the Future* provides new insights into ways of helping both students and teachers think more critically and creatively about their own future and that of the wider society. It acknowledges the crucial role of education in helping young people understand the nature of local and global change and the social and environmental impacts such change will have on their future. Setting out a clear educational rationale for promoting a futures perspective in education, it provides both stimulating and practical examples of futures-orientated classroom activities. It also includes fascinating research into children's views of the future.

Groundbreaking in its fruitful application of ideas from futures studies to the education of teachers and young people, *Lessons for the Future* challenges much professional thinking about the nature and purpose of education, and affirms its role in contributing to a more just and sustainable future.

This book will be of interest to a wide range of educators, whether in relation to teaching, course planning, professional development or institutional policy, both in schools and teacher education. It will be of particular interest to those concerned with PSE, values education, social and environmental issues, and the future of education. It will also provide a key text for those working in fields such as future studies, citizenship education, education for sustainable development and global education.

**David Hicks** is Professor in the School of Education at Bath Spa University College. He has published widely in the fields of geographical education, citizenship education, futures education and global education.

# Futures and Education Series
Edited by Richard A. Slaughter

**New Thinking for a New Millennium**
*Edited by Richard A. Slaughter*

**Educating beyond Violent Futures**
*Francis P. Hutchinson*

**Reframing the Early Childhood Curriculum**
*Jane M. Page*

**Lessons for the Future: the missing dimension in education**
*David Hicks*

# Lessons for the Future

## The Missing Dimension in Education

David Hicks

London and New York

First published 2002
by RoutledgeFalmer
11 New Fetter Lane, London EC4P 4EE

Simultaneously published in the USA and Canada
by RoutledgeFalmer
29 West 35th Street, New York, NY 10001

*RoutledgeFalmer is an imprint of the Taylor & Francis Group*

© 2002 David Hicks

Typeset in Times New Roman by
Keystroke, Jacaranda Lodge, Wolverhampton
Printed and bound in Great Britain by
Antony Rowe Ltd., Wiltshire

All rights reserved. No part of this book may be reprinted
or reproduced or utilised in any form or by any electronic,
mechanical, or other means, now known or hereafter
invented, including photocopying and recording, or in any
information storage or retrieval system, without permission
in writing from the publishers.

*British Library Cataloguing in Publication Data*
A catalogue record for this book is available
from the British Library

*Library of Congress Cataloging in Publication Data*
A catalog record for this book has been requested

ISBN 0-415-27672-1

To my granddaughter Holly and her future

# Contents

|   |   |   |
|---|---|---|
| *List of illustrations* | | ix |
| *Foreword, by Wendell Bell* | | xi |
| *Acknowledgements* | | xvii |
| 1 | Remembering the future: a personal/professional journey | 1 |
| 2 | Reclaiming the future: what every educator needs to know | 11 |
| 3 | A lesson for the future: young people's concerns for tomorrow | 26 |
| 4 | A geography for the future: some classroom activities | 40 |
| 5 | Towards tomorrow: strategies for envisioning the future | 53 |
| 6 | Retrieving the dream: how students envision their preferable futures | 60 |
| 7 | Stories of hope: a response to the psychology of despair | 68 |
| 8 | Always coming home: identifying educators' desirable futures | 78 |
| 9 | Living lightly on the earth: a residential fieldwork experience | 90 |
| 10 | Teaching about global issues: the need for holistic learning | 98 |
| 11 | Questioning the century: shared stories of past, present and future | 109 |
| 12 | Epilogue: some lessons for the future | 122 |
| | *Bibliography* | 131 |
| | *Index* | 140 |

# Illustrations

## Figures

| | | |
|---|---|---|
| 3.1 | Hopes for personal future | 32 |
| 3.2 | Fears for personal future | 32 |
| 3.3 | Hopes for local future | 34 |
| 3.4 | Fears for local future | 34 |
| 3.5 | Hopes for global future | 35 |
| 3.6 | Fears for global future | 35 |
| 3.7 | Optimism over the future | 36 |
| 4.1 | Things that are changing | 43 |
| 4.2 | The extended present | 45 |
| 4.3 | Alternative futures | 46 |
| 4.4 | Four scenarios | |
| | (a) More of the same | 48 |
| | (b) Technological fix | 49 |
| | (c) Edge of disaster | 50 |
| | (d) Sustainable development | 51 |
| 10.1 | Learning about global futures | 102 |

## Tables

| | | |
|---|---|---|
| 4.1 | Hopes and fears for the local area | 41 |
| 4.2 | Hopes and fears for the global future | 42 |
| 6.1 | Elise Boulding's 'baseline' future | 63 |
| 6.2 | Students' preferred futures, 2020 | 65 |
| 6.3 | Evaluation of workshop effectiveness | 66 |
| 7.1 | Sources of hope (1) | 75 |
| 8.1 | Social educators' preferred futures, 2025 | 83 |
| 8.2 | Sources of hope (2) | 84 |
| 11.1 | Preferred futures, 2050 | 118 |
| 11.2 | Sources of hope (3) | 119 |

# Foreword: Preparing for the future

*Wendell Bell*

As we begin a new century we are being told that the human community faces many threats to its future survival and well-being. The possible problems are well known: a polluted environment; climate change; dwindling resources (including the coming scarcity of potable water); deadly local conflicts; religious fanaticism and intolerance; terrorism and violent aggression; continuing illiteracy (there are a billion illiterate people in the world today); unnecessary deaths of people from diseases we could prevent or cure (if we were to deliver modern methods of sanitation and health care to them); the production of weapons of terrible destructive capacity; the violation of human rights in many parts of the world (including the continuing subjugation of women); and increasing inequalities between the affluent and the poor both within and between countries. If we believe these problems are likely to occur, we can easily foresee a future of despair.

Yet that is not the whole story. For the human community is at the same time being told that it faces a dazzling array of opportunities. Such developments include: advances in technologies such as genetic engineering; nanotechnology that may usher in a new post-industrial revolution on a chip, allowing the manufacture of self-replicating 'knowbots' designed to produce goods and services and eliminate many scarcities; the development of new materials, smarter and more durable than those now available; fusion, solar and other energy technologies that will be nearly pollution free and that will be abundant; and information technologies that will transform human communication and make knowledge more accessible to all.

Other possible developments include greater worldwide cooperation, the guarantee of personal security and human rights, and the further spread of democratic participation in collective decision making. They include the suppression of local conflicts, the prevention of violence as a means of resolving disagreements, agreement on global ethics, global systems of governance whilst respecting autonomies of local groups, and a world court of justice. They include opportunities for education and meaningful work regardless of gender, age, race, ethnicity or nationality. And they include, among other things, accelerating exploration of space and, eventually, establishing human settlements on the moon, Mars and elsewhere. Considering these things, we can foresee a sustainable future of hope, individual satisfaction and social harmony.

What will the actual human future turn out to be? Will it be a living chaos of innumerable people consumed with fear and hate, intent on doing violent harm to others and heading toward an apocalyptic end? Or will it be, to the contrary, a win–win world of human cooperation and mutual achievement in which people are forgiving, compassionate, generous, competent and responsible, and in which people will enjoy long and peaceful lives? Or will the human future be something else? Of course, we do not know. But we do know this: whatever the future will be depends importantly on what we humans do. Human life is a series of decisions, the outcome of which will shape our future. Within limits, we create our own individual and collective futures by our acts of commission and omission.

Moreover, we also know that what humans do importantly depends on their images of the future. Such images shape and guide human decisions to act. They contain the world-views that people have, their definitions of the situation and the goals of human endeavor. Images of the future are among the causes of present behavior, as people either try to adapt to what they see coming (such as evacuating an area in advance of a forecast hurricane) or to act in ways that will construct the future that they want (like working to get good grades in secondary school in order to be admitted to a university of their choice). In fact, the fate of entire civilizations – as Frederik Polak has shown in his two-volume work *The Image of the Future* – may rest on whether or not dominant images of the future in a society are, on the one hand, positive and idealistic or, on the other, negative and pessimistic.

Also, what humans decide to do depends on their beliefs about what is possible. Possibilities for the future are real and exist in the present. Alternative possibilities for the future are often unseen or ignored by people as they go through the routines of their everyday lives. As a result, their future may turn out to be considerably less than it might otherwise have been because they fail to ferret out many of the real possibilities and choices that they have. People often limit their own futures, seldom trying anything new, or different, or better. Closing off many of their options, they condemn themselves to the limitations of the present. What the future will be, then, is partly conditional on people thinking creatively, on their ability to seek out and to make visible present possibilities for the future.

Additionally, how people decide to act depends not only on their beliefs about what is possible, but also on their beliefs about what is probable under different conditions. If there were no order and predictability in the world, then it would be impossible to plan and pursue projects to achieve goals. All would be chaos. But despite many random processes, the natural and social worlds are largely ordered in complex and detailed, but often knowable, ways. There are laws and rules, social conventions, cultural customs, habitual routines, and cause and effect. There are schedules for trains and planes, appointments with doctors and dentists, opening and closing times of businesses and institutions, announcements of concerts and elections, and daily and weekly cycles of repetitive events such as commuting between home and work. We know when and where a football game is scheduled to start, even if we do not know which team will win. Although results do not always correspond to expectations, customs and schedules usually give us good guesses about future events. Time moves inexorably on and, under ordinary circumstances, the social order is a more or less stable network of expectations, social coordination and intersecting time trajectories of people, actions and events.

To behave responsibly, intelligently and competently, people need to know with some accuracy the likely future consequences of their own actions and those of others. Otherwise, how can they decide what to do to make their way in the world and achieve their goals? Put more generally, people need to know what the likely future will be, *if* things continue as they are. Or *if* conditions change in particular ways. Or *if* they or others behave in this way or in some other way.

Finally, the future depends on people's ability to judge preferable futures, to understand the human values and goals that define conceptions of the good society. Knowing what is possible and probable helps to guide our actions, but we need to know more. We also need to know what is desirable. Which futures should we want to achieve? Which futures should we try to avoid? To choose to act one way or another involves an evaluation of the desirability of alternative futures. Of all that is possible, what do we want? We may know that we can do X and that doing so will probably achieve Y. But do we really want to achieve Y? If not, then do not do X. In other words, we must ask what kind of future we want.

To find an answer requires moral analysis. In discourse with others, we must judge what is good and what is bad, for ourselves as individuals, for our families and the communities within which we live, for our nations and, indeed, for the entire human community. Some visions of the future, as we saw above, are totally undesirable. By contrast, others are highly desirable and lead to meaningful and satisfying lives. Moral judgments about preferable futures merge with beliefs about possibilities and probabilities and become part of our images of the future. Such images can become the foci of critical discourse, decision making and social action. When they do, they are powerful drivers of social change.

Possible, probable and preferable futures – these are among the important elements of futures thinking. They are what we need to know in order to design effective actions to create the future. We cannot make a conscious decision to act (or not to act) without futures thinking, because the consequences of any contemplated action (or inaction) occur in the future. How well people carry out the tasks of futures thinking largely determines the kind of future that they will have. Futures thinking that is competently done tends to produce beneficial results, while incompetent futures thinking tends to produce undesirable futures.

The interaction of the possible, probable and preferable makes the process of forecasting complicated, because, as people act on their predictions, they may influence events, making the predictions self-altering – either self-fulfilling or self-negating. Thus, many useful predictions are presumptively true at the time they are made, but turn out to be terminally false when the time of the prediction comes about, precisely because the prediction itself led to appropriate action that changed the predicted outcome. For example, the doctor tells you that you will probably develop lung cancer unless you stop smoking. You believe his prognosis and do not want to have lung cancer. Thus, you stop smoking. As a result, you live on without ever getting the disease.

Sophisticated decision makers, of course, take into account this reflexive aspect of people's anticipations, just as they try to foresee the unintended and otherwise unanticipated consequences of all social action. Thus, it is much too simple to evaluate a prediction or a forecast by whether or not it turns out to be true or false in the end. A prediction, once made, can itself become a causal factor in any situation. If people use it as a basis for taking action that negates it, or that fulfills it, a presumptively true prediction may turn out to be terminally false, or a presumptively false prediction may turn out to be terminally true (for example, the classic case of the spread of *false* rumors that a bank will fail resulting in so many people withdrawing their money that it does fail).

Futures thinking, both among ordinary people and even among top government and business leaders as they decide great issues of the day, is often done poorly. The future, thus, is often strewn with shattered hopes and broken dreams. When confronted with complicated life choices, people tend to follow past routines blindly, to fall back on habitual reactions. They are often reluctant to step into the unknown and to explore alternative possibilities, a range of contingent probable outcomes, and their own values about what is really desirable. Even when they do give the future some detailed and conscious thought, they may process information inaccurately and may not reach valid interpretations. Clearly, people need better cognitive maps of the future and guidance about how to use them in order to navigate to the destinations of their choice.

Fortunately, during the last half century, futurists have constructed such maps and provided such guidance, which today are important parts of the field of futures studies. Futurists propose to help people manage the future so as to achieve the futures that they want. They do this by prospective thinking, by attempting to discover or invent, examine and evaluate, and propose possible, probable and preferable futures. They have created a theory of knowledge, a variety of research methods, and a body of systematic principles and empirical research findings to help people make more informed decisions and take more responsible and effective actions to achieve desirable futures.

Long before this new century began futurists felt that the year 2000 could be the marker of a coming sea change. Such works include Daniel Bell's *Toward the Year 2000*, Robert Jungk and Johan Galtung's *Mankind 2000*, and Herman Kahn and Anthony Wiener's *The Year 2000*, all published in the late 1960s. They followed earlier works by pioneering futurists such as Harrison Brown's *The Challenge of Man's Future*, Dennis Gabor's *Inventing the Future*, Bertrand de Jouvenel's *The Art of Conjecture*, and Frederik Polak's *The Image of the Future*.

By the 1970s, futures studies was up and running, the seventies being, perhaps, one of the most significant decades in their development to date. Alvin Toffler published his best-selling *Future Shock* in 1970, spreading the message of accelerating social changes engulfing the world and the need for futures thinking to deal with them. The Club of Rome brought forth *The Limits to Growth* by Donella

Meadows *et al.* in 1972, which, incredibly for a statistical report of computer simulations, sold more than 9 million copies in twenty-nine languages. Although *Limits* was as widely condemned as it was acclaimed, its environmentalist message was heard. Humans, concluded Meadows *et al.*, cannot go on indefinitely using the Earth's limited resources, dumping wastes in the Earth's air, land and oceans, and adding billions more people to the Earth's population without overshoot and collapse.

Daniel Bell, in 1973, produced *The Coming of Post-industrial Society: A Venture in Social Forecasting*, which became one of the most widely cited scholarly books of all time. In it, Bell describes the coming of a service economy, the transition to an information society and to a future-orientated time perspective, and the rise of universities, research organizations, and intellectual institutions as central to the coming society of the future.

Since the 1970s, tens of thousands of articles and books on the future have been published. Futures thinking has become a routine part of the activities of some government agencies, military planning groups and corporate departments. Centers and institutes of futures studies have flourished. Many countries of the world now have a professional national association of futurists, and global organizations also exist, such as the World Future Society and the World Futures Studies Federation. Established journals of futures studies and foresight continue to exist and new ones arrive on the scene.

By the end of the 1990s, futurists were consolidating their past work, building foundations and agenda-setting for new advances in the future. Today, although its boundaries are wide and permeable, futures studies is a distinct discipline. It has its own publications, organizations, theories and methodologies. Its practitioners, increasingly, share conceptual and theoretical commitments, purposes, ethical principles, empirical research and scholarship, professional ideals, a sense of community as futurists, and a growing body of substantive principles and knowledge that can be taught to others, and that can be put to practical use. See, for example, Richard Slaughter's three-volume *The Knowledge Base of Futures Studies*, my two-volume *Foundations of Futures Studies*, Kurian and Molitor's two-volume *Encyclopedia of the Future*, and Schwartz, Leyden and Hyatt's *The Long Boom: A Vision for the Coming Age of Prosperity*.

Despite several hundreds of futures courses now being taught throughout the world and despite a handful of university programs in which students can receive formal degrees in futures studies, there is a glaring imbalance in educational institutions at every stage of learning and in every country. Educational institutions now devote many resources to recovering and preserving the past by studying history, as indeed they should. But they devote relatively few resources to the study of the future.

This imbalance is increasingly dangerous as we enter a century of increasingly rapid change. If we wish to prepare the next generation to deal with the future, if we wish to arm them with the intellectual tools to create desirable futures both for themselves and for their societies, then we ought to establish and expand the systematic study of the future in our schools. Additionally, as complements to existing Departments of History, in all the world's universities we ought to establish Departments of Futures Studies. In them, scholars and scientists of foresight could expand their study of the fan of alternative futures and teach students the purposes, principles, procedures and practices of futures thinking. In the twenty-first century, no one's education should be adequate without a clear futures studies component. We desperately need a future-orientated curriculum at all levels of education.

This is why I am enthusiastic about the publication of *Lessons for the Future: The Missing Dimension in Education* by Professor Hicks. First, David Hicks is the ideal author of such a book. He has been active in the futures community and concerned with education for the future for many years. He has created a body of futures work himself that is well known and highly respected among futurists on several continents. David Hicks is a master of his chosen topic; there are few scholars anywhere in the world as qualified as he is to teach us about 'education for tomorrow'.

Second, *Lessons for the Future* is a work of exceptional quality, a brilliant exemplar and synthesis of futures studies as it bears on education. Selectively focusing on what teachers and students need to know about the study of the future, David Hicks brings together in a clear, accessible and graceful way much of what futurists have learned during the last fifty years. For educators, the book provides a set of concepts and principles that will help them convey to their students the importance of futures thinking to their lives. For students, the book reveals how to explore possible, probable and preferable futures so that they can act now to benefit their future selves. It shows students how to take action that will empower them, now and throughout their lives, to better achieve their life goals and to participate effectively in the processes of community and national decision making.

Readers of *Lessons for the Future* will learn how to deal with the threats to their future, both those they face in their individual lives and those that they confront as members of the global human family. They will learn, too, how to identify and seize the opportunities that exist to create a flourishing personal future for themselves, while at the same time responsibly contributing to a future of freedom and well-being for all humanity.

Wendell Bell<br>
Professor Emeritus of Sociology<br>
Yale University

# Acknowledgements

My thanks are due to many and various people. Four who, in particular, inspired this work were: Dan Dare, 'pilot of the future', in that wonderful childhood comic *The Eagle*; Robin Richardson, who set me on my educational path twenty-five years ago; Elise Boulding and her work on envisioning the future; and Martha Rogers with her research on learning about global futures.

Those who directly assisted me with the research described here were: Cathie Holden (Chapter 3); Kay Wood (Chapter 7); Andy Bord (Chapters 9 and 10); Richard Ward (Chapter 9). Thank you all. I am also indebted to the World Wide Fund for Nature UK which funded my early work on education for the future. I greatly appreciated the enthusiasm of Hywel Evans at RoutledgeFalmer and the US reader who was so supportive of my manuscript.

Fellow futurists by whom I feel especially supported are: Rick Slaughter, Sohail Inayatullah, Ivana Milojevic, Hazel Henderson, Tony Stevenson and Cesar Villanueva. Among futures educators I must thank Frank Hutchinson and Jane Page, as well as all those students who have taken my futures modules at Bath Spa University College. Special friends who have helped me keep going in turbulent times are Patrick Whitaker, John Hammond, Kay Wood and Andy Bord. I must also thank Tyna for the timely blessing that she provided.

Many children, students and teachers have contributed to this book, and in particular I need to thank all those who participated in my Hawkwood conferences in 1996, 1997, 1998 and 1999. Finally, my thanks to all those LEAs and others who have invited me to speak to them and who remind me that this work really *does* make a difference. My thanks to you all.

## Sources

*Chapter 1*: The original version of this first appeared in: R. Zuber, (ed.) (1994) *Journeys in Peace Education: Critical Reflection and Personal Witness*, Peace Education Reports No. 14, University of Malmo, Sweden; reproduced with permission.

*Chapter 2*: Parts of this chapter first appeared in *The Australian Journal of Environmental Education*, vol. 9 (1993), pp. 71–84; reproduced with permission.

*Chapter 3*: The original version of this first appeared in *Futures*, 28 (1), D. Hicks, A lesson for the future: young people's hopes and fears for tomorrow, pp. 1–13, © Elsevier 1996, reproduced with permission from Elsevier Science.

*Chapter 4*: The original version of this first appeared in *Teaching Geography*, 23 (4) (1998), pp. 168–73; reproduced with permission.

*Chapter 5*: The original version of this was first published in *Environmental Education Research*, 2 (1) (1996), pp. 101–8, published by Carfax Publishing, Taylor & Francis Ltd.

*Chapter 6*: The original version of this first appeared in *Futures*, 28 (8), D. Hicks, Retrieving the dream: how students envision their preferable futures, pp. 741–9, © Elsevier 1996, reproduced with permission from Elsevier Science.

*Chapter 7*: The original version of this was first published in *Environmental Education Research*, 4 (2) (1998), pp. 167–78, published by Carfax Publishing, Taylor & Francis Ltd.

*Chapter 8*: The original version of this first appeared in *Futures*, 30 (5), D. Hicks, Always coming home: towards an archaeology of the future, pp. 463–74, © Elsevier 1998, reproduced with permission from Elsevier Science.

*Chapter 9*: The original version of this was co-written with Andy Bord and Richard Ward and first appeared in *The Australian Journal of Environmental Education*, vol. 15/16 (1999/2000), pp. 147–51; reproduced with permission.

*Chapter 10*: This paper was co-written with Andy Bord and first published in *Environmental Education Research*, 7 (4) (2001), pp. 413–25, published by Carfax Publishing, Taylor & Francis Ltd.

*Chapter 11*: The original version of this first appeared in *Futures*, 32 (5), Questioning the millennium: shared stories of past, present and future, pp. 471–85, © Elsevier 2000, reproduced with permission from Elsevier Science.

### Figures

*Figures 3.1, 3.2, 3.3, 3.4, 3.5, 3.6 and 3.7* are reprinted from *Futures*, 28, D. Hicks, A lesson for the future: young people's hopes and fears for tomorrow, pp. 1–13, © Elsevier 1996, with permission from Elsevier Science.

*Figures 4.1, 4.2 and 4.4* are reprinted with permission from D. Hicks, *Citizenship for the Future: A Practical Classroom Guide*, © WWF-UK, 2001.

*Figure 10.1* is reprinted with permission from M. Rogers, *Learning about Global Futures: An Exploration of Learning Processes and Change in Adults*, Ontario Institute for Studies in Education, 1994 (unpublished doctoral thesis).

# Chapter 1

# Remembering the future
## A personal/professional journey

## Summary

How do educators become drawn towards socially critical approaches to education, and what are some of the formative influences that can contribute to such a stance? This opening chapter uses autobiography to explore ways in which personal and professional journeys are often inextricably bound together. First as a geography teacher, and later as a curriculum developer and teacher educator, the author has helped to pioneer innovative and practical ways of teaching about global and futures issues in school and higher education. It is from personal narrative that the substantive concerns of this book unfold and the investigations into futures-orientated teaching that follow can best be understood in this wider life-context.

## Prologue

> Why should we take it for granted that an author's personal feelings and thoughts should be omitted [from a book]. . . . After all, who is the person collecting the evidence, drawing the inferences, and reaching the conclusions? By not insisting on some sort of accountability, our academic publications reinforce the third-person, passive voice as the standard, which gives more weight to abstract and categorical knowledge than to the direct testimony of personal narrative and the first-person voice.
>
> (Ellis and Bochner 2000: 734)

A spring day in the early twenty-first century in north-east Somerset. My usual journey to work. As I pass through the university gates and look at the great trees, the fields and the stream in the valley, I give thanks, as always, to those whose hands laid out this beautiful place in time past. At lunchtime I walk around the lake and reflect on how I came to be in this place. A series of 'snapshots' crosses my mind's eye.

- A small baby being carried downstairs by his father in the middle of the night. He is laid on a table and wrapped in a green check blanket. The house is full

- of fear as the air-raid siren wails in the adjacent street. The fear of being bombed, of being buried alive in an East Anglian town in 1943.
- A young boy in a classroom being chastised by a teacher for talking. He remembers the angry women who punished him at primary school. Now it is aggressive male teachers at this English grammar school who offer him violence as well. He is constantly in trouble, lives in fear of many lessons.
- A student sitting in his small study at a college of education in London. He is surrounded by friends who are trying to persuade him to come out for a drink. What he actually wants is to be left in peace to meditate. They are intrigued by, but make fun of, his interest in Eastern philosophy.
- A classroom in Gloucestershire on a sunny spring day. A young teacher stands at the blackboard teaching about current trouble spots in the world. On the wall a map of the Middle East and pictures from a colour magazine about the war in Vietnam. A pupil complains, 'This isn't geography!'
- A young family is encamped with hundreds of others on the coast of Scotland with one of the Greenpeace vessels standing offshore. These activists have come from all over the country to protest at the construction of a nuclear power station and to support those preparing to occupy the site.
- A room full of teachers reflecting on what global education means for them. They are working in small groups, sharing their experiences, as part of a professional development programme in a faculty of education at an Australian university. The facilitator moves from group to group, moved by their commitment and enthusiasm.
- A summer school for teachers in Nova Scotia. Participants have come from many parts of the province to explore ways of putting a practical futures dimension into their work. The week is intense and many feel transformed by the experience. The convenor feels that the facilitator wrought some kind of 'magic' with the group.

If I had glimpsed any of these scenes before they actually occurred, my response would have been one of surprise. Yet looking backwards, they make sense, part of my thread of becoming. I can see how each scene was enfolded in what came before, like a seed awaiting the right time to fruit, my personal and professional lives always inextricably intertwined.

In considering the 'genres of ethnography' available to qualitative researchers today, Tedlock (2000) draws attention to the valuable contribution that autobiography and personal narrative can make to writing and research. The writer's own experience thus becomes a topic for investigation in its own right (Denzin 1997), challenging positivist notions of the possibility of there ever being a 'disengaged observer'. Behar (1997) and others have stressed the importance of 'knowing who the author is' and the need therefore to reflect our vulnerabilities in our texts. Feminist researchers have also stressed the importance of incorporating personal experiences and standpoints in academic work, often by starting with a story about themselves (Reinharz 1992). Ellis and Bochner (2000: 746) comment that 'A text that functions as an agent of self-discovery or self-creation, for the author as well as for those who read and engage the text, is only threatening under a narrow definition of social inquiry, one that eschews a social science with a moral center and heart'. This chapter thus situates the research that follows within the context of the author's own personal and professional journey. It invites the reader to become a co-participant – emotionally, morally and intellectually – in both the life story and the analysis that follows.

## Learning to resist

I was born during the darkest days of World War Two in Ipswich, the county town of Suffolk. It must have been an awful time for my parents, my twin and I imbibing their fear and anxiety even before we were born. I arrived prematurely and my brother died soon after birth. My first clear memories as a small child are of the early 1940s – being carried down into the air-raid shelter in the garden during the war. I was quite clear what I felt about school after my first morning there. I declared it was nice but that I didn't need to go back in the afternoon. The problem, as the teachers saw it, was that I was too talkative. This constantly got me into trouble with the women teachers, who tried to shout, bellow and slap me into submission. I put both my anarchist tendencies and my strong sense of injustice down to those childhood experiences at primary school.

Grammar school was even worse. Here several of the male teachers regularly tried to threaten, humiliate and beat me into submission. I learned to survive through solidarity with my peers and a well-developed sense of humour nourished by the Goon Show. Not until I was 16 did teachers leave me in peace to enjoy my education. In particular, I grew to love geography and English literature, mainly because of particular teachers who really knew their craft well. So I developed a passion for this world, finding out about how it was formed and how it worked. I loved doing fieldwork and getting to know an area of countryside well.

The first turning point came in the summer when I left school at 18. Walking with my friend Patrick in the country we pondered the meaning of life, whether God and evil existed or not. I think we imagined we would find the answers to such questions in just a few months. I had been brought up as a member of the Church of England but now found such Christianity far too narrow and hypocritical.

## Quest for meaning

I left school in 1960 to read geography at the University of Exeter. Devon was a revelation of beauty with its hills and woods, and I lived on a farm in the Exe valley. Appalled at what I read about the superpower arms race and the consequences of nuclear war, I joined the Campaign for Nuclear Disarmament. As a result of my nuclear nightmares, from which there seemed no escape, I fell into deep despair and disillusionment. The second turning point came with Resnais's documentary *Night and Fog*, a graphic and harrowing portrayal of the Nazi concentration camps. As I watched the bulldozers pushing piles of emaciated bodies into mass graves, I felt as if I too had gone over the edge into the abyss.

Cycling home that night I was overcome by tears of despair but, even as I wept, I realised that the only thing anyone *could* do was somehow to work towards making the world a better place. It was a simple, perhaps even simplistic, realisation but a powerful one. My sense of purpose and vocation in this life dates from then. My mid-term exam results were bad, however, and I spent the rest of that year sitting in the library reading modern English literature, philosophy and comparative religion.

I then spent three years at Borough Road College in London, now part of Brunel University. The urban scene, suburbia, the traffic hurt my soul after the beauties of Devon, but I studied geography and English there, training to be a secondary teacher. I also immersed myself in comparative religion, especially Buddhism, Zen and Taoism. I left college in 1964 clear that I wanted to use my geography teaching to help students make sense of important issues in the world around them.

## Education and action

My first teaching post was at a rural secondary school just outside Gloucester. I was fascinated at this time by books that gave potted histories of world trouble spots. Lessons on issues such as the Arab–Israeli conflict, the Vietnam war, the nuclear arms race, and issues of global wealth and poverty became an ongoing element in my teaching. In the mid-1960s it was still unusual for anything to be taught about global issues. So I gained a lot of confidence in my subject and learned my craft as a geographer in the Severn valley, the Cotswold Hills and the Forest of Dean. I also ran the Duke of Edinburgh's Award Scheme and enjoyed teaching map reading and camping skills and taking groups on expeditions. I gained my Mountain Leadership Certificate so I could take groups to really wild areas, teaching them to take an interest in the terrain they traversed, to feel at home on the fells, to be safe in all weather conditions. I could now pass on my own love of the countryside, hills and mountains to others.

From a small rural school I went to a larger urban one as a Head of Department. Again I was able to indulge my curriculum interests, and produced a syllabus that included orienteering in Thetford Chase and the study of local and global issues, but other interests were beginning to blossom too. It began with a TV documentary in the early 1970s on environmental issues, called 'Tomorrow has been cancelled due to lack of interest'. It wasn't going to be if I had anything to do with it. A colleague and I set up a local environmental action group – one of the first in East Anglia. We campaigned on a wide range of issues, from recycling and pollution of the local river to waste dumping and urban redevelopment. It gave me a taste for campaigning, organising, engaging in social and political action, working directly to make my bit of the world a better place. This, in turn, led to my increasing politicisation as I became interested in grassroots activism and the history and philosophy of anarchism. My heart still leaps when I read:

> There is nothing integral to the nature of human social organisation that makes hierarchy, centralisation and elitism inescapable. These organisational forms persist, in part because they serve the interests of those at the top. They persist, too, because we have learned to accept roles of leadership and followership ... even the eradication of coercive institutions will not automatically create a liberatory society. We create that society by building new institutions, by changing the character of our social relationships, by changing ourselves – and throughout that process by changing the distribution of power in society.
>
> (Ehrlich and Ehrlich 1979)

During my last year of teaching in school I came across the work of the World Studies Project run by Robin Richardson. I was impressed by his experiential and participatory approach to learning, his resources for teachers and, especially, the workshops he ran. Three of the chief influences on his work were Carl Rogers, Paulo Freire and Johan Galtung. At a conference Robin organised in 1974, called 'Only One Earth,' I remember thinking that if I had any ambition in life it was to be actively involved with such events myself.

After ten years of teaching in schools I wanted a change, and it came with a post at Charlotte Mason College of Education in the Lake District. Here I was the geographer in the Environmental Studies department and the Social Studies team. It was at this time that I began writing, initially about the courses I was teaching (Hicks 1976, 1978). Soon after arriving in Cumbria I become concerned about the nuclear waste reprocessing plant at Windscale and, at an anti-nuclear meeting in Lancaster, I met Paul Smoker who ran the Peace and Conflict Programme at the university. When my contract ended at the college in the mid-1970s I arranged to do an MSc with him.

For this I decided to investigate the extent to which a global perspective was found within initial teacher education in the United Kingdom. To my excitement, I found that all sorts of interesting things were going on, with tutors often drawing on the work of development educators and the World Studies Project. Both approaches proposed a socially critical and person-centred pedagogy closely allied to that of peace education. In 1977 I attended the International Peace Research Association conference in Stockholm, where, for the first time, I had to face the questions then being raised by the women's movement. This began a long and often painful process of examining my own oppressive role in supporting patriarchal structures. My MSc was completed in the same year, but I decided to stay registered for my PhD while also becoming the first Education Officer for the Minority Rights Group in London. This took me into the field of multicultural education and caused me to write *Minorities: A Teacher's Resource Book for the Multi-ethnic Curriculum* (Hicks 1981). In particular, I became interested in the issue of racist bias in teaching materials, a topic which geographers had rarely touched on (Hicks 1980; Walford 2001).

During this period I also continued my environmental activism. Plans were afoot for a major expansion of the Windscale plant and thousands of signatures were collected and delivered to the Department of the Environment in London. On the way back we decided to set up a Network for Nuclear Concern to link anti-nuclear groups in north-west England. Groups were far flung and often suspicious of possible attempts to organise them, but the Network came to play an important role in coordinating opposition to the expansion and presenting evidence at a public inquiry. They were powerful days, appealing in particular to my anarchist and oppositional tendencies. I realise now how the faceless men of British Nuclear Fuels felt just like those adults who had oppressed me as a child. They too were threatening my existence. So we organised, networked, campaigned, learned about non-violent direct action, and suffered the inevitable burnout.

## Teacher education

As a result of my research I had become inspired by the pedagogy of peace education and wished to promote this within teacher education. I secured a base for this at St Martin's College of Higher Education in Lancaster (now St Martin's University College) and launched two major initiatives, the Centre for Peace Studies and the World Studies 8–13 project. During its nine-year lifespan the Centre was a unique initiative in the United Kingdom. Its aims were (1) to promote within education awareness of issues relating to peace and conflict; (2) to interpret and clarify the existing educational responses to such issues namely, education for peace, world studies and development education; (3) to identify the priorities for curriculum innovation in these fields at both primary and secondary level.

When the Centre first opened in 1980, interest in education for peace *per se* among teachers was negligible. However, escalation of the nuclear arms race saw a growing debate about what teachers should, or should not, be teaching in the classroom. Much of the Centre's work focused on giving educators a much broader definition of peace and conflict, from the personal and local to the national and global. In 1984 and in 1986, the International Year of Peace, I was invited to Australia by John Fien, a geographical educator from Brisbane. On both occasions I lectured and ran workshops in several states and found a flourishing interest in peace education. I was impressed by the work of people such as Toh Swee-Hin and Frank Hutchinson. John's interests were in geography, environmental and social studies and he played an important part in radicalising these curriculum areas (Fien and Gerber 1988).

At this time I served on and advised several Local Education Authority working parties on peace education, including those for Manchester and Lancashire. Important links were made with issues of gender and race at that time by Betty Reardon (1985) in her book *Sexism and the War System*. Attacks from the political Right, however, began to increase in intensity. They argued that 'peace studies' was an initiative from the Left, backed by the Campaign for Nuclear Disarmament, bent on indoctrinating pupils into anti-nuclear activities. Much of my time was spent assuring teachers, school governors and others that this was *not* in fact the case. An important outcome of this work was the first UK handbook on peace education (Hicks 1988).

Visits to Canada in 1986 and 1988 led to further contacts with peace educators from North America, but I realised that peace education in the United Kingdom now lagged behind developments in Canada and Australia. Also around this time I met Joanna Macy and became very interested in her work on despair and empowerment (Macy and Brown 1998). By 1989, however, the educational scene had drastically changed. Continued attacks on humanistic and radical approaches to teaching, coupled with the introduction of Thatcher's national curriculum, resulted in a shift towards both more formal methods and a more standardised curriculum. Peace education *per se* became totally marginalised within schools.

A national curriculum project called World Studies 8–13 was also set up in 1980 and directed by myself and Simon Fisher (who had taken over from Robin Richardson at the World Studies Project). World studies was a shorthand term used in the United Kingdom to refer to the need for a global perspective in the curriculum. (Internationally, the term 'global education' is more commonly used.) Initially, the project worked with pilot schools in Cumbria and Avon, developing teaching materials and refining a participatory and experiential approach to in-service work. Eventually the project had a network of contacts in fifty LEAs (that is, half those in England and Wales), using as their key resource *World Studies 8–13: A Teacher's Handbook* (Fisher and Hicks 1985).

Visits to Italy, Australia and Canada proved that world studies approaches were popular in other educational contexts as well. At this time the political Right launched forceful attacks against both world studies and peace education. As the Conservative government's new national curriculum began to tighten its hold, the project's second publication for teachers, *Making Global Connections*, came out (Hicks and Steiner 1989). The project moved to Manchester Metropolitan University in 1989 (Steiner 1993, 1996) and, now renamed the Global Teacher Project, it is currently based at Leeds Metropolitan University.

World Studies 8–13 was one of the most innovative curriculum projects of the 1980s, and its success lay in the principles and procedures that it used. It showed how local and global issues were related, it acknowledged the experience and expertise of teachers and worked with them in a person-centred way. It produced materials that were practical and fun to use, it excited both teachers and pupils, it made learning interesting and dealt with issues of immediate interest in the real world. My own interests were increasingly focused on the need for a clearer futures dimension in the curriculum as well as a global one.

## Alternative futures

My life changed dramatically as I came to understand through therapy the depth of childhood wounds and the ways in which they affect adult life. I also became interested in eco-feminist perspectives (Shiva 1989; Diamond and Orenstein 1990) and ventures which attempted to draw the personal and political together. I didn't like them being seen as separate, for my experience was that they were intertwined. Starhawk's (1990) *Dreaming the Dark*, which drew together issues of spirituality, sexuality, community and non-violent direct action, served to reinforce what I so strongly felt. I was also interested in the role of new social movements in creating change and the suggestion that we might be witnessing a major shift of paradigm in the Western world (Milbrath 1989).

In 1989 I returned to London to start a new project at the University of London Institute of Education and to join a course on Facilitator Styles at the University of Surrey. I wanted to deepen my knowledge of group facilitation and the course at Guildford seemed an excellent opportunity to do this. It combined practical training in facilitation skills with a wide-ranging exploration of different approaches to therapy and counselling (Heron 1989). It was exhilarating and exhausting, but by the end of it I was much clearer about who I was and what my skills were.

At the Institute of Education I set up the Futures Project with funding from the World Wide Fund for Nature UK. World studies argued that the spatial dimension in the curriculum emphasised the local and national at the expense of the global, and it stressed the need to explore the interrelationships *between* local, national and global. On the temporal dimension the past and present are given more attention than the future, creating a 'temporal imbalance' in the curriculum. The question I thus put to teachers is 'If all education is for the future where is the future explored in education?' The early days of the project involved exploring the futures field to see what would be of direct use to teachers (see Chapter 2). My particular interest is in how to explore the nature of more just and sustainable futures. But to do this teachers and pupils need to develop a new vocabulary of futures-orientated thinking.

My interest in drawing together the personal, political and spiritual was reinforced by working with Joanna Macy and radical theologian Matthew Fox in 1991. I was one of a group of facilitators for an event entitled 'Death and Resurrection of Self, Society and World'. Joanna spoke of facing our pain and despair in the face of planetary issues. Matthew spoke of cosmology, our spiritual traditions and the need to confront injustice (Fox 1994). I felt I was coming full-circle in some way, that my interests were approaching some sort of synthesis.

In the early 1990s I left London to live in Bath and moved the project to Bath Spa University College. It gradually became more difficult to work with teachers, due to the growing financial and curriculum restraints on schools, but I was able to visit Italy and Canada and greatly enjoyed running summer schools with David Ferns in Nova Scotia. When funding for the Futures Project ended, I became a member of staff at the university. Here, in the School of Education, I am able to share my interests through teaching modules on education for change, radical education, education and environment, education for the future and citizenship education. My teaching and research interests arising from these fields are reflected in the following chapters.

## Connecting the threads

What sense do I make of this personal and professional journey? What are the connecting threads and how are they related? My basic life-pattern was laid down during my childhood and adolescence in the 1940s and 1950s. My experiences at school led to a heightened sense of injustice and a dislike of mindless authority. My Church of England up-bringing and my parents' conservatism led me to reject traditional explanations of life, both religious and political. Above all, in my youth I learned about *resistance*. I learned to be critical, to keep asking the awkward questions. And through enjoying geography at school I learned to love this planet.

As a young student this lead to an ongoing search for life's meaning, a *spiritual* quest that preoccupied me for most of the 1960s. Faced with the evils that I witnessed around me, I could see no other course of action than to try and make the world in some small way a better place. Herein lay the seeds of my later concern for justice, equality and peace. I became a geography teacher, and taught my students both to love the land and to explore local and global issues.

The 1970s saw my move into teacher education and post-graduate research. In particular it was a decade of *politicisation* for me through involvement with the environmental, the anti-nuclear and the women's movements. I began writing and broadened my geographical interests to embrace environmental studies, world studies, development education and multicultural education.

The 1980s was a period of synthesis, putting global education and peace education on the national map in a practical and creative way. It was also a time of *personal* growth as I learned about my self through individual and group therapy. My spiritual interests resurfaced via eco-feminism and the renaissance of Goddess spirituality. Increasingly I felt aware of the links between the personal, political, spiritual and planetary.

In the 1990s I focused less on problems and more on the directions and *visions* that we need in order to create a more just and sustainable future. Joanna Macy's deep ecology and Matthew Fox's creation spirituality combine many of my major concerns. Changing the world and changing oneself always go hand in hand. Personal and political equity and justice can never be separated, even if many would wish them so.

What sustains me are my hopes and dreams for a more just and sustainable future, although, as the anarchist slogan reminds us, the revolution is the journey not an end event. I am always interested in interfaces, the personal and political, the spiritual and planetary, the inner and outer. This is where growth occurs as we struggle to create synthesis. The dominant social paradigm in the Western world, with its mechanistic viewpoint, has taught us to see things as separate and unconnected. The emerging holistic paradigm, with its ecocentric perspective, argues that we must now start putting things back together again. The journey for the new century must be one from fragmentation to wholeness, in ourselves and our relationships, both with each other and with non-human species. It is about ending the dis-membering and beginning the re-membering. It is a journey from personal and global separation and alienation towards true justice and community.

My purpose in this introductory chapter has been to create a 'conversation' with the reader by using personal narrative to invite identification with the existential dilemmas I faced and my personal/professional attempts to resolve these. It also helps situate the wider text socially and historically, thereby allowing a deeper engagement with the educational issues raised. But, as Lincoln and Denzin (2000: 1049) argue:

> Of the following, we continue to be certain: the qualitative researcher is not an objective, authoritative, politically neutral observer standing outside and above the text . . . qualitative inquiry is properly conceptualised as a civic, collaborative project. This joins the researcher and researched in an ongoing moral dialogue.

Over a hundred years ago, in the 1890s, William Morris, the great designer, writer and political activist, struggled with his socialist comrades in Britain to create a better world – one which, in part, we have now inherited. Much of his passion was inspired by the long utopian tradition which, 'at its most radical, invades the prevailing concept of reality, undermines certainties about what humans must always be like, and casts doubt upon the inevitabilities of the relations of everyday life' (Coleman and O'Sullivan 1990). At heart, it is this 'imagination' which has sustained my work over the last four decades and from which this book has emerged. Here, in this new century, it seems increasingly important that we too should try to live as if we were experiments from that future which we so desire.

Chapter 2

# Reclaiming the future
## What every educator needs to know

### Summary

Students in education now will become the parents, voters, workers, business people and visionaries of the early twenty-first century. How are they being prepared for these responsibilities and how are they being helped to think more critically and creatively about the future? This chapter sets out the need for a more futures-orientated approach to education, arguing that the future is often a missing dimension within the curriculum. An essential element of responsible citizenship has to be the ability to consider different scenarios for the future, whether personal, local or global, which will lead to more just and sustainable ways of living.

### These times

Dickens's description of the 1790s in *A Tale of Two Cities* could also well be a description of the early years of the twenty-first century:

> It was the best of times, it was the worst of times, it was the age of wisdom, it was the age of foolishness, it was the epoch of belief, it was the epoch of incredulity, it was the season of Light, it was the season of Darkness, it was the spring of hope, it was the winter of despair, we had everything before us, we had nothing before us.

In 1992, just before the Earth Summit in Rio, the *Guardian* ran the headline 'Earth's future in the balance, say top scientists'. This referred to a joint report from the US National Academy of Sciences and the 'normally-conservative' Royal Society of London, which warned of the catastrophic consequences of resource consumption, environmental destruction and population growth.

> In essence, the two bodies – to which both governments have always turned for advice on scientific matters – have endorsed many of the warnings issued over the past 20 years by pressure groups such as Friends of the Earth and Greenpeace, making it difficult for politicians to say scientific opinion is divided, or that talk of global disaster is scaremongering.
>
> (*Guardian* 1992: 1)

Some thirty years ago, the 'limits to growth' debate was sparked off by publication of the first report from the Club of Rome. The authors of that report, which predicted global collapse if present trends continued, also published a sequel called *Beyond the Limits* (Meadows *et al.* 1992). While they argue that the world has already overshot some limits, they also show that a sustainable future *is* technically and economically feasible. When and where in our teaching do we encourage students to debate such issues?

Each year, the Worldwatch Institute in Washington, DC, publishes two seminal reports highlighting contemporary global issues and debates. These are *State of the World* (Brown *et al.* 2001), which examines progress being made towards a more sustainable society, and *Vital Signs* (Brown *et al.* 2001b), which explores key environmental and social trends that will shape the future. Some current trends at the start of the new century are: (1) world grain harvests no longer keep up with population growth; (2) leading energy corporations shift investment from oil, coal and nuclear to wind and solar power; (3) the world is getting warmer – with the fourteen warmest years since record-keeping began in 1866 all occurring since 1979; (4) educational levels rising worldwide and in particular an increase in the education of women in developing countries; (5) a wired world with 100 million people on-line, nearly all in North America, Europe, Australia and Japan, however; (6) after peaking in 1984, global military expenditure has dropped by nearly 40 per cent; (7) environmentally, an accelerated loss of forest, plants and animal species. Each of these trends will have differing impacts regionally and locally, now and in the future.

The Real World Coalition, in *The Politics of the Real World*, writes about how people viewed the future in the mid-1990s:

> The year 2000 approaches. The symbolism is irresistible. A new century, a new millennium: a time to take stock of the past, to look forward to the future with hope and purpose . . . [but] for large numbers in Britain today, the new century is not a source of hope. The predominant mood, if anything, is of fear. People are anxious about the future, about the world they are leaving their children.
>
> (Real World Coalition 1996: 1)

There is also reference to a public opinion poll, which asked people what they felt about the future. Only 12 per cent thought the future would be better than today, 60 per cent felt it would be worse, and 25 per cent said about the same. These figures well stand for the early twenty-first century too.

If young people are to become effective citizens of tomorrow, it is vital that the curriculum contains both a global dimension, which explores the multiple spatial interrelationships between local, national and global communities, and a futures dimension, which similarly looks at the temporal interrelationships between past, present and future. Where in education might this occur?

## New movements

Some years ago a colleague was showing a French visitor around his institution as they discussed the state of education. 'Where', asked his guest, 'are the new utopians to be found?' 'Well actually', responded my colleague, 'I'll be seeing them at a meeting tomorrow.' He was in fact referring to a group of people who were all, in different ways, involved in the promotion and practice of socially critical education via a range of issue-focused approaches to teaching and learning which had emerged internationally in the 1970s and 1980s. These include global education, development education, environmental education, futures education, peace education and other related fields. Each in different ways is concerned about the 'state of the planet'.

Clearly, many of these initiatives overlap, although each begins with a different substantive concern. One thing they share in common is the belief that, in a democratic society, young people should have the opportunity to explore a range of contemporary issues, whether to do with development, interdependence, peace and conflict, human rights, the environment, or race and gender. Ian Lister (1987) noted these common concerns and saw the proponents of such 'new movements' as vanguard educators in social and political education. What these initiatives also have in common is an interest in the human condition and its improvement. As such, they all involve teaching for a better world and thus implicitly, if not explicitly, are concerned about the future.

As well as being issue-orientated, these new movements also share a common concern for the *process* of education, and it is this process which is at the root of effective learning. How young people learn is as important as what they learn, and most practitioners in these fields put great emphasis on participatory and experiential modes of learning which foster both pupil autonomy and the development of critical thinking skills. Effective learning is seen as arising out of affirmation of each pupil's individual worth, the development of a wide range of cooperative skills, the ability to discuss and debate issues, to reflect critically on everyday life and events in the wider world, and to act as responsible citizens. This requires a teaching style which is open and facilitative. Its intention is to model participation, cooperation and justice in everyday classroom interaction, believing that the medium of learning should match the message.

This interest in, and emphasis on, process and methodology places particular emphasis on exploring the relationship between the personal and the political. It draws on two long-standing traditions in education. The first focuses on the person and thus on the wholeness of the individual; the second on the political and the wholeness of society. As Richardson argues:

> Both traditions are concerned with wholeness and holistic thinking, but neither, arguably, is complete without the other. There cannot be wholeness in individuals independently of strenuous attempts to heal rifts and contradictions in wider society and in the education system. Conversely, political struggle

to create wholeness in society – that is, equality and justice in dealings and relationships between social classes, between countries, between ethnic groups, between women and men – is doomed to no more than partial success and hollow victories, at best, if it is not accompanied by, and if it does not in its turn strengthen and sustain, the search for wholeness and integration in individuals.

(Richardson 1990: 6–7)

This is the fertile ground in which a critical education for the future can grow. It is ground which has been well prepared, both theoretically and practically, over the past decades. In particular, proponents of global education have the interest, skills and expertise that is needed to bring the futures dimension more alive in the curriculum. If the task in the 1980s was to emphasise the need for a global dimension in the curriculum, the task for the new century is to highlight also the need for a clear futures dimension.

## A futures dimension

In his classic work, *Learning for Tomorrow: The Role of the Future in Education*, Toffler writes:

All education springs from images of the future and all education creates images of the future. Thus all education, whether so intended or not, is a preparation for the future. Unless we understand the future for which we are preparing we may do tragic damage to those we teach.

(Toffler 1974: cover)

Most curricula, whether explicitly stated or not, see school as having a significant role to play in preparing pupils for the opportunities, responsibilities and experiences of adult life. One might therefore expect exploration of the future – personal, local, global – to be an integral part of young people's studies.

The future is an integral part of everyday life. We spend a large part of our time thinking about it. Identifying goals for the future enhances our ability to work in the present, adds to our motivation and helps give us direction. While on the one hand the future is intangible, it is also of crucial importance; human existence cannot be conceived of without it. Yet because there are no facts about the future we often neglect it or leave it to others: management, economists, politicians, multinational corporations, and it's *their* future that we finish up with. We often know what we don't want in the future, but may be less clear about what we *do* want. We are also more used to thinking about personal rather than societal or global futures, but the times require that we think much more explicitly about the future and about how local and global futures are interrelated. In particular, we need to ask the question: where are we going and where do we want to go, locally, nationally and globally?

In fact, there is no such thing as *the* future (singular), for at any given moment in time any number of futures (plural) are possible. The term 'alternative futures' or simply 'futures' is often used as a shorthand reminder of this.

For an exploration of alternative futures to be of use it needs to be remembered that different people and groups have quite different views of the future: a middle-class child in Perth, a homeless woman in London, an unemployed worker in Dresden or a logger in Brazil. Clearly, some groups in society also have much more power than others to define the future, generally those who are rich in the global system or who wield power through, say, transnational corporations, international banking, governments, the military, the media. In some sense such groups also colonise the future, particularly big business with its constant creation of new 'needs' for tomorrow.

What we expect in the future is often not what we would wish for. Helping pupils identify both their probable and preferable futures is an essential element in preparation for adult life. Teaching only about problems can alienate young people. Learning about possible alternatives and proposed solutions is much more likely to motivate them. The current state of the planet requires that exploration of just and sustainable futures becomes a major priority at all levels of society. If we can work to envision such futures with others, then we stand a better chance of bringing them about. Future generations would surely ask no less of us than this. We live in times of uncertainty and surprise. 'Now' is the place where we need to anticipate and prepare for the future before it unfolds.

The purpose of education for the future is to enable pupils to explore various scenarios which may emerge from current trends and to explore the implications of these. Once we realise that we cannot opt out of the future it takes on a radically new meaning. All actions and choices, including choices *not* to act or choose, have future consequences. Slaughter writes:

> We cannot alter the past, but we have common interests in achieving life-preserving, sustainable futures. It follows that a central task for teachers is to explore with their pupils some of the major problems and possibilities that lie ahead and thereby sensitize them to the implications of choices and actions in the present.
>
> (Slaughter 1985: 5)

Education for the future helps to clarify the range of alternatives that lie before us in any situation, giving rise to a wider choice of options, which in turn leads to more thoughtful and responsible action in the present. The rationale for including a clearer futures dimension in the curriculum is shown in the box on pp. 16–17.

Discussion with teachers and teacher educators in various countries reveals, however, that the future is largely a missing dimension within education. Gough's (1990) investigation of the portrayal of futures in educational discourse is invaluable. As a result of his exploration of educational documents Gough was able to identify

RATIONALE FOR A FUTURES DIMENSION IN THE CURRICULUM

*Pupil motivation*

Pupil expectation about the future can affect behaviour in the present – e.g., that something is, or is not, worth working for. Clear images of desired personal goals can help stimulate motivation and achievement.

*Anticipating change*

Anticipatory skills and flexibility of mind are important in times of rapid change. Such skills enable pupils to deal more effectively with uncertainty and to initiate, rather than merely respond to, change.

*Critical thinking*

In weighing up information, considering trends and imagining alternatives, pupils will need to exercise reflective and critical thinking. This is often triggered by realising the contradictions between how the world is now and how one would like it to be.

*Clarifying values*

All images of the future are underpinned by differing value assumptions about human nature and society. In a democratic society, pupils need to be able to begin to identify such value judgements before they can themselves make appropriate choices between alternatives.

*Decision making*

Becoming more aware of trends and events which are likely to influence one's future, and investigating the possible consequences of one's actions on others in the future, leads to more thoughtful decision making in the present.

*Creative imagination*

One faculty that can contribute to, and that is particularly enhanced by, designing alternative futures is that of the creative imagination. Both this *and* critical thinking are needed to envision a range of preferable futures, from the personal to the global.

> *A better world*
>
> It is important in a democratic society that pupils develop their sense of vision, particularly in relation to more just and sustainable futures. Such forward-looking thinking is an essential ingredient in both the preserving and improving of society.
>
> *Responsible citizenship*
>
> Critical participation in democratic life leads to the development of political skills and thus more active and responsible citizenship. Future generations are then more likely to benefit, rather than lose, from decisions made today.

common types of reference to the future – tacit, token and taken for granted. *Tacit futures* are all those which are assumed and never brought out into the open. They remain hidden and unexplicated but nevertheless present. Thus the future may not even be mentioned in an educational document but assumptions about it are still tacitly present. *Token futures* often involve clichés and stereotypes presented in a rhetorical fashion. Gough (1990: 303) notes, 'When one finds "the future" (or a futures-oriented inference) in the title of an educational document it usually means much less than might be expected.' *Taken-for-granted futures* occur whenever a particular future, or range of futures, is described as if there were no other alternatives. Discussion of the future framed solely in terms of science and technology or work and leisure would be in this category.

Explicit study of the future does, however, have long-standing international and academic credentials. Most recently the field of futures studies has been mapped in depth by Slaughter (1996) in *The Knowledge Base of Futures Studies* and by Bell (1997) in *Foundations of Futures Studies*. Detailed educational case studies from the United Kingdom, Canada, the United States and Australia can be found in *Futures Education: The World Yearbook of Education 1998* (Hicks and Slaughter 1998).

The particular aims of futures education are to help teachers and pupils

- develop a more *future-orientated* perspective both on their own lives and on events in the wider world;
- identify and envisage *alternative futures* which are just and sustainable;
- exercise *critical thinking* skills and the *creative imagination* more effectively;
- participate in more thoughtful and informed *decision making* in the present;
- engage in active and *responsible citizenship*, both in the local, national and global community, and on behalf of present and future generations.

Aims such as these are of interest to a wide range of educators concerned with subjects such as English, maths, science, technology, geography, history, modern languages, business studies and religious education. They are also of particular relevance to equal opportunities, multicultural education, and cross-curricular themes such as environmental education, citizenship, and personal and social education. Some of the themes that can contribute to a futures dimension in the curriculum are shown in the box below.

---

SOME KEY FUTURES THEMES

*Where are we now?* (**the situation**)

*Thinking about the future*   Students share their thoughts about the future, about change and about time. They then use these ideas to reassess their current interests and concerns, in relation to their personal lives, the local community and the wider world.

*Probable futures*   Next, students clarify how they expect the future to be. What different futures do they anticipate for themselves and for the world? What images do others have of such probable futures? What do they feel about these probable futures?

*Where do we want to get to?* (**the vision**)

*Preferable futures*   Students identify their own preferred futures and compare them with those of others. In particular, they are encouraged to explore what just and ecologically sustainable futures might look like, both globally and in their own daily lives.

*The utopian tradition*   Students study ways in which people in the past have tried to put their visions of the 'good society' into practice. By exploring life in selected utopian communities and initiatives, pupils further clarify their own values and priorities for the future.

*How do we get there?* (**action**)

*Projects for change*   Students then re-focus on the present by investigating the work of groups and organisations committed to creating critical social change; e.g., in relation to justice, the environment, equal opportunities and development issues.

*Active citizenship* Students then plan and carry out their own projects for change at home, in school and the local community. In so doing they learn to match future local/global needs with socially critical and ecologically responsible citizenship today.

## Further themes

*Views of the future* Students need to understand that people's hopes and fears for the future vary enormously, depending, among other things, on age, gender, race and class. These differences require further investigation. Any notion of a just world order requires that we ask, '*Whose* future are we talking about? Who will benefit/lose from such a scenario?'

*Feminist perspectives* Some of the most important insights into preferred futures come from feminist writers, whether in relation to development issues, the environment or utopias. These perspectives are crucial to any understanding of the present and how it has been shaped by patriarchy. Attention needs to be paid to how such insights can be used to challenge and transform current education practice.

*Inter-generational justice* Notions of justice used by philosophers, lawyers and economists now include the rights of future generations. Justice thus has both a spatial and a temporal dimension. Students therefore need to explore what rights future generations might wish to have and how they, as the present generation, might help to safeguard these.

*Envisioning the future* Images of the future are a crucial element in the creation of change. They play a powerful role in what students think is, or is not, worth doing. Envisioning preferable futures needs to be a vital part of futures education and much more common in educational practice generally. This needs to acknowledge both our pain for what is happening to this planet and its peoples and also our visions of more just and sustainable futures.

*Sustainable futures* One of the most important concepts for students to understand and explore is the notion of sustainable development. This emphasises the need to meet the needs of the least advantaged in society, without causing further environmental damage and also taking into account the needs of future generations. Students need to understand what sustainability looks like at home, in school and in the local community, and be prepared to commit themselves to its furtherance.

> *Utopian traditions* There is a long utopian tradition in Western thought which focuses on the nature of the good life and the perfect society. Utopians may present their ideas as fiction, where the ideal society is set in some other time or place, or as a programme for political action, or in the form of intentional communities. This living tradition has much to offer students and others interested in exploring preferable futures.

## Children's responses

Children are always immediately interested in the future, as these quotations from 10-year-olds in a Berkshire school indicate:

- 'I hope that in the future there will be no more war and hunger and the world will become green and everybody will care to make it better and the world will unite again.'
- 'My hopes for the future are to have a happy life and to live for a long time and have a nice family. And that my cat and goldfish will live for a long time.'
- 'One of my fears for the future is that I'll get hijacked in a plane, another is that I'll get stuck in an elevator.'
- 'Another world fear is that the atmosphere will get too polluted so that we cannot live any longer.'

Concerns range from the personal to the global, from families and pets to the welfare of other people and the planet itself.

One way of beginning work on futures with pupils is to encourage them to ask questions about the future (Hicks 2001). This activity permits them to formulate their *own* questions about the future and the teacher to identify the particular interests of the class. The procedure is as follows:

- Each student writes down individually five questions they would like to ask about the future. There is no limitation as to their scope. They may be personal, about school, the local community or events in the wider world.
- Students then work in groups of five or six to put similar questions together and then rank them by popularity. Each group can thus identify their 'top ten' questions.
- Alternatively, the aim can be to arrive at a list of top ten questions for the class as a whole. One such list eventually agreed upon by a class of 7-year-olds in Taunton is shown below.
- Students then work in pairs to discuss and research possible answers to their questions, using resource books in class or in the library and/or by interviewing other people in school or at home. Of course, in many cases they will find that

there is no agreed 'right' answer and thus begin to learn about the need to tolerate ambiguity and differences of opinion.
* Finally each pair presents to the class in written, visual or spoken form the results of their research and/or interviews.

## *7-year-olds' questions about the future*

* Will I still be laughing when I'm 50?
* Will we find out about Pluto?
* Will our friends be the same?
* What will our mums and dads look like?
* Will there be a war soon?
* How many more animals will be extinct?
* Will the jungles be destroyed?
* Will cars still have leaded petrol?
* Will pollution stop divers going under the sea?
* Will people still use pencils?
* Will they invent irons to run off solar heat?

Year 3, Blackbrook School, Taunton

## Views of the future

One of the most useful distinctions that futurists make is between probable and preferable futures. Probable futures are all those which seem *likely* to come about. They involve the projection of current trends – for example, in relation to economic growth, unemployment or global warming – and making forecasts about what is therefore expected to happen. Preferable futures are all those which one feels *ought* to come about to achieve a particular set of value preferences. This involves envisaging, for example, what a more just and ecologically sustainable community or world might look like and the steps needed to bring it about.

Images of the future play a crucial role in the development of human life and culture. Contemporary society cannot be explained simply as the result of the *push* of the past; it is also deeply influenced by the *pull* of the future, offering new and exciting possibilities. As Slaughter writes:

> Images of the future present us with options and possibilities from which we can select and choose or with which we may argue and debate. Either way, they are active, shaping components of human consciousness. The main purpose of considering futures, and images of futures, is not to predict what will happen in any hard or precise sense . . . it is . . . to discern *the wider ground from which images are constituted* so as to take an active part both in creating and nurturing those which seem worthwhile.

(Slaughter 1991: 499–500)

In the mid-1950s Fred Polak, the Dutch sociologist, wrote a major study of images of the future in Western society, putting forward the thesis that certain images of the future develop an unusual potency and act as a societal time-bomb, creating what he called a 'breach in time'. Radically new visions of the future, he argued, can produce a sharp temporal and historical discontinuity (Polak 1972). Faced with new and potent images of the possible future, society begins to mobilise its creative energies in response.

However, Polak also argued that a lack of guiding images can lead to a loss of direction and purpose in society. In particular, he argued that there was a decline of imaging capacity in Western society, that in the mid-twentieth century compelling positive images of the future were few and far between. The experiences of World War One, World War Two, and then the advent of the Cold War, made the future not a place of hope but one of fear. The long tradition of utopia gave way to one of anti-utopia, or dystopia, exemplified by Huxley's *Brave New World* and Orwell's *Nineteen Eighty-four*. It is almost impossible to watch television or go to the cinema without being faced by images of disaster, breakdown and despair. At some deep level we are haunted by these themes yet sense that something is missing. What is missing are positive guiding images of the future which can give both direction and the confidence that things *can* be radically different.

What the futures field can offer is a wealth of expertise in exploring more desirable futures. It is as important to know where we *want* to get to as where we do not. As Richardson commented:

> A map without utopia on it, it has been said, is not worth consulting. . . . Admittedly there are disadvantages in dreams and ideals, the disadvantages of unreality and abstractions. But frequently it also clears and strengthens your mind if you venture to dream for a while, as concretely and as practically as possible, about the ideal situation to which all your current efforts are, you hope, directed.
>
> (Richardson, cited in Hicks and Steiner 1989: 10)

This brings us back to Polak's thesis that without strong, positive guiding images of the future a society loses its directions. We now need to be clear about what we *want* in the future as well as what we oppose.

Every culture has its image, whether religious or secular, of a past or future better world: Eden, Arcadia, the Golden Age, Paradise. In Western thought this aspiration became known as 'utopia', a term coined by Thomas More in 1516 and based on a pun in Greek: eutopia, the good place, was also utopia, no place. Utopias provide blueprints for a better future society and may be presented as fiction, an ideal society set in some other time or place, or as a programme for political action and change. Utopias always have a double-edged message: a critique of present imperfections and a vision of a better world. Literary utopias range from Plato's *Republic* to William Morris's famous *News from Nowhere* and Charlotte Gilman's *Herland*. Various groups in Europe, North America and Australia have planned and set up their own utopian communities with varying degrees of success. In Britain, for example, there were Gerrard Winstanley and the Diggers in the seventeenth century, religious groups such as the Shakers in the eighteenth century, and secular communities such as Robert Owen's New Harmony in the nineteenth.

As Krishan Kumar (1991: 3) writes, 'Utopia's value lies not in its relation to present practice but in its relation to a possible future. Its "practical" use is to overstep the immediate reality to depict a condition whose clear desirability draws us on, like a magnet.' The utopian tradition, both as literature and as lived experience, has constantly inspired critical action for change and can provide a rich source of nourishment for the creative imagination today.

We live in one world that contains many other voices telling of the future, voices that the dominant culture may often ignore. Here are four such voices.

From Yothu Yindi's album, *Tribal Voice*:

> Well I heard it on the radio/ And I saw it on the television/ Back in 1988, all those talking politicians/ Words are easy, words are cheap/ Much cheaper than our priceless land/ But promises can disappear/ Just like writing in the sand/ This land was never given up/ This land was never bought and sold/ The planting of the Union Jack/ Never changed our law at all/ Now two rivers run their course/ Separated for so long/ I'm dreaming of a brighter day/ When the waters will be one.
>
> (Yothu Yindi 1992)

Eduardo Galeano, the Uruguayan writer:

> There is just one place where yesterday and today meet, recognise each other and embrace, and that place is tomorrow. Certain voices from the American past, long past, sound very futuristic. For example, the ancient voice that still tells us we are children of the earth . . . [and] that speaks to us of community heralds another world as well. Community – the communal mode of production and life – is the oldest of American traditions. . . . It belongs to the earliest days and the first people, but it also belongs to the times ahead and anticipates a New World. For there is nothing less alien to these lands of ours than socialism. Capitalism, on the other hand, is foreign: like smallpox, like the flu. It came from abroad.
>
> (Galeano 1991: 103)

Vandana Shiva, on women, ecology and development:

> Seen from the experience of Third World women, the modes of thinking and action that pass for science and development, respectively, are not universal and humanly inclusive, as they are made out to be; modern science and development are projects of male, western origin, both historically and ideologically. . . . The industrial revolution converted economics from the prudent

management of resources for sustenance and basic needs satisfaction into a process of commodity production for profit maximisation. . . . The new relationship of man's domination and mastery over nature was thus also associated with new patterns of domination and mastery over women, and their exclusion from participation as partners in both science and development.

(Shiva 1989: 56)

And Eleonora Masini, on women as builders of the future:

Women are better adapted for the change from the industrial society to a new society, because women are not the carriers of the values of the preceding industrial society. As they were not the builders of the future in the preceding society, they may become the builders of the future in a different society. As they were invisible in the industrial society, women may become visible and constructive in a post-industrial society.

(Masini 1987: 434)

When and where in our teaching do we encourage students to listen to such voices and to draw on such traditions?

Traditional models of development and ideas of progress narrowly focus on economic growth (GNP as a measure of consumption), with its intrinsic discounting of other 'costs'. Thus some people benefit at the expense of others, people benefit at the expense of the environment and people today benefit at the expense of future generations. Such development is not sustainable because globally more people are getting poorer, finite resources are diminishing and the environment is beginning to suffer irreversible damage. As Postel comments:

Put simply, the global economy is rigged against both poverty alleviation and environmental protection. Treating the earth's ecological ills as separate from issues of debt, trade, inequality, and consumption is like trying to treat heart disease without addressing a patient's obesity and high cholesterol diet: there is no chance of lasting success.

(Postel 1992: 5)

Working towards a sustainable future requires production planned to meet human needs together with a more just distribution of resources (Real World Coalition 2001). It means reducing the harmful effects of industry and new technology, challenging company policies which are dangerous to people and the environment, stopping aid programmes which are inappropriate and damaging, reducing overconsumption and waste, restraining population growth, distinguishing clearly between wants and needs, and organising locally, nationally and internationally for appropriate change. When and where do we encourage our students to explore the need for, and nature of, sustainable development in both the local and global community?

This is where we need to begin, with a socially critical approach to learning that recognises the importance of changing both self and society. It requires 'the courage to admit and bear the pain of the present world' while at the same time 'keeping a steady eye on [our] vision of a better future' (Meadows *et al*. 1992: 230).

Chapter 3

# A lesson for the future
## Young people's concerns for tomorrow

**Summary**

This chapter explores the importance of research into popular images of the future. It begins by reviewing the scattered literature on this in relation to both adults' and young people's views of the future, and summarises what is known so far. It then describes the findings of a research project which set out to update the UK data in this field. Working with pupils from a variety of schools in the south-west, this examined how pupils in the 7 to 18 age range see the future at personal, local and global scales. The responses of different age groups are examined as well as variations based on gender. Finally, the significance and importance of such research for futurists and educators generally is emphasised.

Until relatively recently futurists appear to have shown only an occasional interest in how young people view the future. Yet, because children will become the adult citizens of the twenty-first century, it is the images that they have now which will influence their aspirations for that future. It would seem imperative, therefore, that more attention be paid both to the images of the future that young people already have and to the sort of education that will promote creative exploration of futures issues. This is a task which must begin with the students themselves and where *they* feel they stand in relation to the future.

That educators are beginning to engage in this process is shown by work emanating from the United Kingdom, Sweden, Australia, Canada and the United States (Longstreet and Shane 1993; Vilgot 1995; Hutchinson 1996; Hicks and Slaughter 1998; Page 2000). Although the word 'future' seldom occurs in UK curriculum documents, teachers nevertheless have a statutory responsibility to prepare students for the opportunities, responsibilities and experiences of adult life. Their adult lives will stretch well into the latter part of the century. Whether they are aware of it or not, all teachers are therefore involved in education for the future.

## Images of the future

Although images of the future are considered to be a central concern of futurists (Huber 1978; Boulding 1988), research on views of the future has been relatively sporadic. Livingstone comments:

> The general importance of images of the future as mediating factors in social action has been postulated . . . by several scholars. Such theoretical work has not distinguished very clearly between hopes (what people want to see) and expectations (what they think will probably happen), or between people's attitudes regarding their personal future and their views on the societal future.
> (Livingstone 1976: 195)

The research described in this chapter adds to what is already known internationally about young people's images. In order to set this study in context, the literature on both adults' and young people's views of the future is sampled first, and some of the main findings outlined.

An interesting South African study by Danziger (1963) in the early 1960s invited participants to write 'histories of the future'. They were asked to imagine themselves as historians looking back on the years 1960–2010 and to describe the major events of that period. From this Danziger found that different social groups had quite different views of the future, ranging from what he called 'conservative' and 'liberal' to 'catastrophist' and 'revolutionist'. He does not say whether any respondents foresaw black majority rule within that period. Cantril (1965) was responsible for a series of studies which enabled him to compare changing hopes and fears for the future. Participants were asked to describe the best and worst possible future they could imagine in ten to twenty years' time. This was then used to establish the upper and lower limits of a ten-point scale. Individuals were then asked to rate themselves and their country on this scale five years ago, in the present and in five years' time.

One of the most authoritative investigations, carried out in the early 1970s, was a comparative ten-nation study entitled *Images of the World in the Year 2000* (Ornauer *et al*. 1976). This major survey involved researchers in Britain, Norway, the Netherlands, Spain, Poland, Czechoslovakia, Yugoslavia, India and Japan. The questions were wide ranging, as the following examples indicate:

- Would you say that you think very much, much, little or not at all about the future of your country, not in a couple of years but, say, in the year 2000?
- How often would you say that you talk with somebody about the future of your country or the world?
- What do you think will be the difference between the year 2000 and today?
- What do you think will be the situation in your country by the year 2000? Do you think that people will be more happy or less happy than they are today?

In general, the tendency to think about the future was found not to be very well developed among respondents. Images of the future most often focused on likely developments in science and technology or problems to do with war and peace rather than broader social futures. By and large, pessimistic visions of the future were better developed than optimistic ones. In the richer nations there was a certain scepticism about science, while in the poorer nations scientific development in any field was generally appreciated. In his conclusion Galtung noted that,

> For the nations in our sample the future seems somehow to be synonymous with a technological future. The future is seen in technical terms, not in terms of culture, human enrichment, social equality, social justice, or in terms of international affairs. . . . People may also think in terms of social future but regard it as unchangeable. But it seems more probable that they have only been trained to think technologically and have no other type of thoughts as a response to the stimulus 'future'; or at least have not been trained to express any other thoughts. And this will then become self-reinforcing since no one will be stimulated by others to think about social futures.
> 
> (Galtung 1976: 56–7)

Soon after this, Livingstone (1983) carried out a similar survey in Ontario, Canada. He was particularly concerned about the gap between the views of the future held by intellectuals and the general public. The majority of his respondents indicated frequent thinking about the future which suggested a shift in interest since the earlier World Images 2000 Project. Whereas fewer respondents claimed any great clarity over their images of the future, corporate capitalists, managers and professional employees claimed the most. The unemployed expressed the greatest future disorientation. A further change noted by Livingstone was an increased likelihood that participants thought about the future in terms of social issues as well as in relation to science and technology.

No similar large-scale projects have been carried out since these two studies. Perhaps this is because, as Moll (1991) has argued, the neo-conservative atmosphere of the 1980s was unsupportive to future-orientated studies. In popular terms, images of the future in the Western world continue to be largely pessimistic. Everitt, writing about the impact of the millennium on the West's imagination, observes:

> At the last *fin de siècle*, the key to the future was held by the likes of Sherlock Holmes. The application of scientific methods to the problems of the present would allow for a benign outcome in the future. As we approach not just another end of century but the end of a millennium . . . our equivalents of the sleuth of Baker Street are Judge Dredd, Robocop, Blade Runner and Terminator, mythic figures in whom futuristic technology and medieval visions of hell have come together to form a nightmarish anti-utopia. We seem to be moving towards 2000 with a head full of fears, to be drifting into a cyber-space populated by monsters of the deep.
> 
> (Everitt 1995: 5)

## Young people's views

Some of the early studies on young people's views of the future feel dated and not much more than attitude surveys (Gillespie and Allport 1955). In the 1970s, Toffler (1974) was one of the first to report a dissonance between what young people expected in their personal lives and how they saw national and global futures. In their own lives they expected to secure a job, get married and raise a family, while in the wider world they foresaw revolution, natural disaster and nuclear war.

> Yet no matter how turbulent a world they pictured, no matter how many new technologies might appear or what political revolutions might take place, the way of life foreseen for themselves as individuals seldom differed from the way of life possible in the present and actually lived by many today. It is as though they believed that everything happening outside one's life simply by-passes the individual. The respondents, in short, made no provision for change in themselves, no provision for adaptation to a world exploding with change.
> (Toffler 1974: 11)

Brown's (1984) study in the mid-1980s is one of the few to have focused on young people in the United Kingdom. In particular, she argued that children's views reflect the social, cultural and political concerns of the times. Initially, 250 16- to 18-year-olds wrote short essays about a day in the year 2000 and the sort of future they would like.

> From these essays certain common themes emerged. Of life in the future as expected: violence, unemployment, high technology, boredom, inflation, poverty, pollution, material prosperity, and, mainly from secondary modern girls, a life not much different from that of today. Of the future as desired, world peace was the most frequently mentioned ideal, and came into the vast majority of essays.
> (Brown 1984: 306)

A large majority of the essays envisaged a highly technological future, although many students seemed uneasy about this. Non-materialistic values were expressed far more often than materialist ones and a 'disenchantment with the modern, consumer society seemed to run through many of the essays' (1984: 309).

In her work as a classroom teacher, Holden (1989) was able to observe the responses of younger children to learning about the future. Having asked her class of 9–10-year-olds to draw their probable and preferable timelines for the future, she comments:

> The children's perceptions of a preferable future indicated preoccupations which surprised me and to which we were to return again and again. . . . Although they did want 'new toys', 'no telling off' and 'different sweets', they also showed a common concern for an end to poverty, crime and war.

> Interestingly ... none expected nuclear war in their own lifetime. ... The fact that all the timelines featured an end to war in their preferable futures, does show children's great concern in this area – nine-year olds are not as innocent as some would have us believe.
>
> (Holden 1989: 6)

Johnson (1987) describes a major survey in the United States which explored pupils' views on personal, national and global futures. Most had a conventional, but optimistic, view of their own personal futures. The majority expected to be married with children, to own a home and a car, to be richer than their parents, and happier than they are now. Both boys and girls showed an increase in awareness with age of the need for gender equity.

Students were less optimistic about the future of the United States, more than half believing that drug abuse and crime would increase. Most felt that the social and economic situation would be the same or worse in the future. Globally, 60 per cent felt that the danger of nuclear war would increase and almost as many saw the depletion of resources and pollution getting worse. The only hope they saw was for an improvement in race relations.

The Henley Centre's (1991) survey of 10–14-year-olds had an environmental focus and revealed a high level of interest, with global issues being cited more often than local ones. Overall children's concerns fell into three broad groups: global issues such as deforestation, the ozone layer, global warming; issues closer to home such as litter, pollution, car emissions; and issues related to animals, especially vivisection and endangered species. The solution to such problems, respondents felt, was a modification of present patterns of consumption rather than any major change in Western lifestyle. All of their drawings of the future were pessimistic.

Hutchinson's (1996) doctoral thesis marks a revival of interest by educators in young people's views of the future. He used a questionnaire with 650 Australian teenagers and follow-up discussion in small groups with a sample of his respondents. He graphically describes their fears for the future:

> Many of the young people in this study expressed a strong sense of negativity, helplessness, despondency and even anguish about the anticipated problems facing their society and the world at large. For a majority, negative imagery of the future ranged from perceptions of intensifying pressure and competition in schools in the twenty-first century to worsening trends in physical violence and war, joblessness and poverty, destructive technology and environmental degradation.
>
> (Hutchinson 1996: 72–3)

Among these concerns it was possible to identify six major themes: (1) an uncompassionate world, depersonalised and uncaring; (2) a physically violent world, with a high likelihood of war occurring; (3) a divided world, between the 'haves' and the 'have nots'; (4) a mechanised world, of violent technological change; (5) an environmentally unsustainable world, with continued degradation of the biosphere; (6) a politically corrupt and deceitful world, where voting is a waste of time.

Hutchinson (1996) was equally concerned, however, to explore young people's preferable futures and found these fell into four broad categories: (1) technocratic dreaming, in which students uncritically accept technofix solutions for all problems (most popular among boys); (2) a demilitarisation and greening of science and technology, to meet genuine human needs; (3) intergenerational equity, accepting responsibilities also for the needs of future generations; (4) making peace with people and planet, via a reconceptualisation of both ethics and lifestyles.

Young people's views of the future reflect the socio-political concerns of the time. Whereas earlier surveys showed optimism over personal futures but pessimism over national/global futures, this dissonance disappears in more recent research. Indeed, Eckersley (1994: 2) notes that 'young people's sense of futurelessness has not lessened with the end of the cold war. Rather the studies suggest that a deepening concern not only about war, but also global environmental destruction, growing violence and inequality, and an increasingly dehumanised, machine dominated world'.

## A UK survey

This survey was carried out with nearly 400 children from schools in the south-west of England. The aim was to discover what young people in the 1990s were thinking about the future. It was the first such survey in the United Kingdom since Brown's (1984) and was innovative in three main ways. First, it was concerned with younger children as well as teenagers; second, it took a developmental perspective across a ten-year age span; third, it explored gender as a key variable. To ensure a representative sample, schools were chosen from both urban and rural environments as well as a variety of socio-economic backgrounds. The children were from four age groups: 7 and 11 (primary school), 14 and 18 (secondary school).

The study was based on a questionnaire, with follow-up discussion with sample groups of pupils. The format of the questionnaire was partly influenced by Ornauer *et al.* (1976). In particular, it focused on the saliency of futures thinking, hopes and fears for the future, and optimism and pessimism over particular issues. While Johnson (1987) looked at personal, national and global futures, we chose local futures rather than national because this was more appropriate for the younger children. The 7-year-olds answered fewer questions than other pupils (thus the gap in some figures below). The survey also looked into the role of school in making students aware of futures-related issues. This chapter reports, in particular, on what was found out about young people's hopes and fears. A more detailed account of the research findings can be found in Hicks and Holden (1995).

How often do young people think and talk about their personal futures? Just under half of the primary pupils claim to think often or very often about this, but only a third say they talk about it with their friends. These figures increase at secondary level, with three-quarters saying they think often or very often about their future and just under half discussing this with their peers. As life becomes more complex in adolescence, there is more to think and talk about. It is interesting to note, however, the gap between inner reflection and the willingness to share this with friends.

So what is it that young people are busy thinking about in relation to their own lives? Figures 3.1 and 3.2 below show their main personal hopes and fears for the

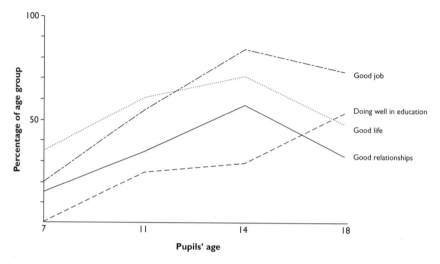

*Figure 3.1* Hopes for personal future

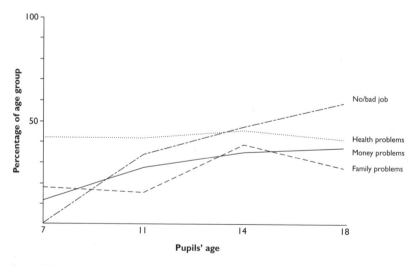

*Figure 3.2* Fears for personal future

future and how these change with age. As is often the case, hopes and fears mirror each other quite closely.

Pupils' main hopes relate to education, work, relationships and achieving a good life. They are what one might expect from children still at school. These hopes all become more important with age and peak at 14, with the exception of doing well in education. It appears that 14-year-olds are more preoccupied with getting a job, doing well in life and their relationships, than older students. Many, of course, will be leaving school at 16. For those who stay on, doing well in exams becomes their main priority.

Students' fears for their own futures again reflect the concerns of today: work, health, money, family. These fears all become greater with age, although concern about health and family problems peak at 14. Fear over not getting a job (or having a job one doesn't like) and having financial problems continues to increase for the oldest age group. It would appear from the above that hopes and fears are strongest around 14. This may either be related to the stage of adolescence or the fact that many will be leaving school in a year or so. Older students share the same concerns but not so acutely, possibly because they have begun to come to terms a little more with these issues or because examinations now take precedence over everything else.

How often do young people think and talk about the future of their local community? The figures for this do not change greatly with age. Most claim not to think about the future of their local area often or very often. Just under two-thirds say that they sometimes think about it, and nearly a quarter say that they never think about the local area in this way. Even fewer talk about it with their friends; half say they never do and 44 per cent say they sometimes do. This would appear to indicate a major lack of interest in the future of their own communities. However, when asked to write about their hopes and fears for the local future their concerns begin to come alive.

The main hopes for where they live are greater prosperity, less crime, better amenities and less pollution. Their fears for the future are over increases in crime, unemployment, pollution and a worsening environment (see Figures 3.3 and 3.4). These hopes and fears show interesting variations with age. The hope for less pollution drops dramatically with age and may well be a reflection of greater realism about this issue. As pupils get older, they increasingly appreciate the need for the local community to be more prosperous. They also worry more about crime, whether vandalism, mugging or rape. The environment generally becomes less of a concern in the face of what appear to be other more pressing issues.

So the local community, whether inner city, leafy suburb or market town, has become a problematic place. The home area is no longer somewhere to feel safe or secure. Amenities, whether for shopping or leisure, are often considered inadequate or under threat. The safety of the streets has become questionable. Young people fear both for themselves and for others, commenting on both the begging and homelessness around them. They also see trees and parks, open space and farmland constantly being threatened.

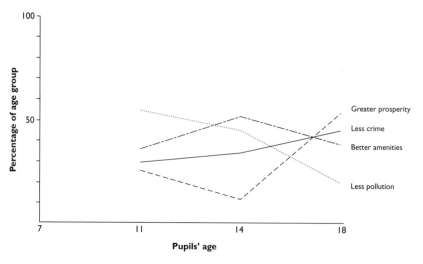

*Figure 3.3* Hopes for local future

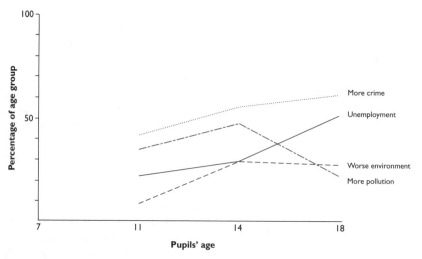

*Figure 3.4* Fears for local future

How often do young people think and talk about global futures? Over half say that they think often or very often about this. Two-thirds say they often or very often see programmes about this on TV, yet over half say they only talk sometimes about this with their friends, and nearly a quarter say they never do. This level of interest is almost as high as that for personal futures but again discussion about these issues is much less.

Young people's hopes and fears for the future of the world show some interesting trends (see Figures 3.5 and 3.6). What is most striking is the degree of concern about war, which, whether expressed as a hope or a fear, greatly outranks all other issues. This concern begins towards the end of primary school and then steadily increases. While the environment is still a major concern, interest peaks at 14 and then begins to decline as it also did in relation to the local area. Interest in issues to do with world poverty and food remain fairly constant, while recognition of the importance of good relationships between countries begins to grow with older pupils. In Figure 3.6 it is interesting to see how fear of disasters, very high with the youngest children, falls dramatically away by the end of primary school.

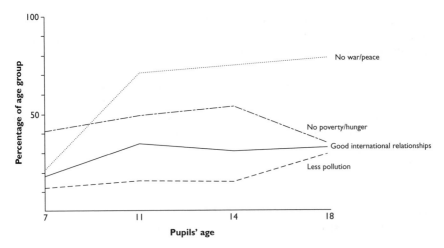

*Figure 3.5* Hopes for global future

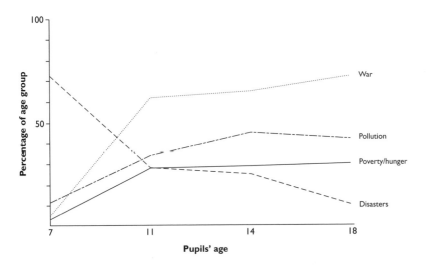

*Figure 3.6* Fears for global future

In relation to their personal futures, optimism decreases somewhat with age: 60 per cent of 7-year-olds think that their life will be much better in the future as against 20 per cent of the 14-year-olds. However, pupils are generally fairly optimistic about their own futures. At primary level, three-quarters think that the future will be a bit or much better than today. At secondary level, nearly 70 per cent still believe this. Young people at school are very aware that most of their life lies before them.

Concerning the future of their local area, a third overall expect this to be much the same. Just over 40 per cent expect it to get better and just under a quarter expect it to get worse. So, after some initial concern, there is a general feeling as children get older that things should improve in the local community.

Optimism over the global future decreases with age. The 7-year-olds are the most optimistic of all. Overall, 42 per cent think that the world will be a better place in the future, a quarter think it will be the same, and a third think that the future will be worse. With nearly 60 per cent therefore thinking that the global future will be the same or worse, this is a rather pessimistic view of the planetary future.

Very little of the existing research on views of the future has paid attention to gender as a variable, a major omission on the part of futurists. When the degree of interest in the future and the degree of hopes are looked at, for example, some clear differences emerge between boys and girls. Consider, for example, the saliency of interest in the future as indicated by thinking and talking about this with friends.

Here girls are more likely than boys to think about the future at all three scales, particularly at primary school and at 18. Girls, however, are much more likely to talk about these matters with each other than boys, particularly their personal futures but also the global future. Girls are also more optimistic about their personal

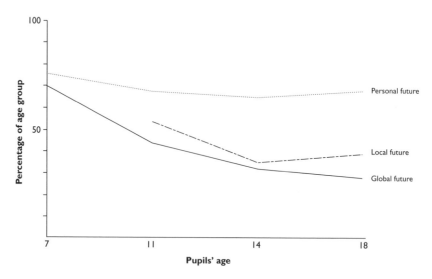

*Figure 3.7* Optimism over the future

futures than boys, except at 14. At secondary school they are less optimistic than boys both about their local area and the global future.

Interesting gender differences can also be observed in relation to hopes for the future. At the personal level girls are much more concerned about education than boys, while the latter are much more interested in the good life. Up to age 14 relationships are much more important to girls than the boys. Boys are more concerned about pollution, both locally and globally, than girls are. After 11 girls are more concerned about relationships between countries than boys.

These differences highlight some interesting avenues for further research. They would seem to reflect some of the gender differences found more broadly within society; for example, a greater sensitivity on the part of women to both relationships and social issues. Thus overall girls show most interest in the future and are more likely to talk about this with others. They value education more and material things less than boys. They are less optimistic about the future – maybe more realistic – than boys, who readily get into the glamour of hi-tech solutions for everything.

Respondents were asked finally how much they had learned about global issues in school and how important they felt this was. Only the youngest age group felt they had learned a lot, which at 7 is probably true. Taking the 11–18-year-olds as a whole, nearly three-quarters felt that they'd learned little or nothing about global issues at school. If this is true it is a worrying comment on their educational experience. Any curriculum which fails to offer a global perspective runs the danger of being merely nationalistic. On the other hand, when asked how important they felt it was to learn about global issues in school, 98 per cent of pupils said that they thought it important or very important.

## Lessons for the future

The survey reported on here supports many of the earlier findings in the literature and also adds significantly to our understanding of young people's hopes and fears for the future. Earlier work suggested that adults' images of the future are not well developed and more likely to be pessimistic than optimistic. It also indicated that people often feel they have very little control over the future, a future more often than not seen as controlled by science and technology. The research on young people's views of the future suggested that they show high levels of concern about contemporary issues and are often fearful about what the future will bring. There has often been a dissonance between their expectations of a fulfilling personal future and a fraught national or global future.

This research confirms that young people's ability to think about the future is not very well developed and that their images tend to be pessimistic. It confirms their concern about current issues and shows that these change depending on the current global situation. Thus war is now their main concern, with environmental issues slipping into second place. The dissonance between personal and global probable futures has disappeared. All levels of society are now seen as problematic.

More recent work in this field is beginning to add further depth to what we know about young people's views of the future. Gidley (1998), while noting the pessimism often found among Australian youth, found that Steiner-educated secondary students had a somewhat different perspective. While they also shared grave concerns about the future of the environment, social justice and conflict/violence, they felt much more proactive about the future. Seventy per cent of these students felt they could make a personal contribution to resolving local problems in some way. It would thus appear that the educational/spiritual philosophy of schools may affect students' own sense of empowerment. Gidley notes that:

> Steiner education provides artistic, imaginative, values-based, meaningful educational experiences and processes which are a counter-balance to the often fragmented, abstract, violent, meaningless and pessimistic messages of our culture provided through the media.
>
> (Gidley 1998: 405)

Eckersley (1999) reports on a major survey carried out by the Australian Science, Technology and Engineering Council, which used a series of scenario-development workshops to explore what young people aged between 15 and 24 expect and want of their society in 2010. Optimism was more common among the younger age group, and men were more optimistic than women about Australia's future. Most do not expect life to be better in 2010. 'They see a society driven by greed; they want one motivated by generosity. Their dreams . . . are of a society that places less emphasis on the individual, material wealth and competition, and more on community and family, the environment and cooperation' (Eckersley 1999: 77). These are themes we will meet again in Chapter 6.

Considerably more work needs to be done by futurists on the significance of images of the future. The studies referred to here clearly show the tenor of such images. If Polak's (1972) thesis is correct, that a society's images of the future reflect its inner health and well-being, then this is indeed a crucial area for research. The West's repertoire of positive future images would seem to be at an all-time low (Everitt 1995), and this vicious circle somehow needs to be broken.

This chapter began by noting a growing interest in the need for futures education in schools. The existing research on children's views of the future is one of the strongest endorsements of that need. Not only is a futures dimension often still missing in the curriculum but so also is a global perspective, two essential prerequisites for an effective education. What is more disturbing is the degree of concern that young people express about the future and the inability of most teachers to respond to this.

However, appropriate responses are available, as shown in the work of O'Rourke (1994), Hutchinson (1996), Rogers (1998) and others. Not only do such educators offer an appropriate curriculum framework for futures education but they also pay due attention to working with students' hopes and fears. In particular, attention is given to exploration of their preferable futures for the local and global community.

This, together with case studies of social change and stories of personal empowerment, allows students to see themselves as potentially proactive rather than merely reactive to change. One process that can be used to facilitate clearer visualisation of appropriate directions for change is the futures workshop and, as I argue later in Chapter 5, this can profitably be used by educators to explore the nature of more just and sustainable futures.

More broadly, radical environmental educators are beginning to point out the role that education itself plays in creating unsustainable futures (Smith 1992; Bowers 1993; Sterling 2001). Thus, while one may start with students' hopes and fears, the trail leads much deeper. Although futures workshops offer one process for counteracting fears and liberating more positive visions of the future, on their own they are still insufficient. Such processes need to be part of an explicit futures dimension at all levels of the curriculum. However, the questions posed by critical futures and environmental educators are now challenging the very nature of education itself. As Orr writes:

> Education in the modern world was designed to further the conquest of nature and the industrialisation of the planet. It tended to produce unbalanced, underdimensioned people tailored to fit the modern economy. Postmodern education must have a different agenda, one designed to heal, connect, liberate, empower, create and celebrate. Postmodern education must be life-centred.
> (Orr 1992: x)

Present and future generations could surely ask no less of educators today than that they should continue in this struggle to create a more life-centred education.

Chapter 4

# A geography for the future
## Some classroom activities

### Summary

Geography as a subject has drawn on many of the insights of global, development and environmental education in order to promote concerns such as global understanding, citizenship in an interconnected world, participation and responsibility, respect for diversity of cultures and the need for a more sustainable world. This chapter illustrates how geographers' interests in the impact of change on people and places has also led to an embracing of many of the insights offered by futures education.

### Teaching for tomorrow

> In urging that we teach a geography of the future, I do not mean to say that we should give up teaching the geography of the past: but we should make that past the servant of the future. If the future is unavoidable, let us at least not walk backwards into it.
>
> (Walford 1984: 208)

A geography of the past or a geography of the future: which shall we choose in school? Backward-looking geography deals with the issues and problems of the late twentieth century, the agenda that teachers themselves grew up with. Forward-looking geography explores the issues and dilemmas of the early twenty-first century, the agenda that our students will have to live with. Will we choose the old century or the new?

Should the task of meeting such needs seem difficult, geographers might recall the key role that exploration played in the evolution of their subject during the nineteenth century. Early geographers often gained their credentials as explorers mapping unknown territories. Such *spatial* exploration played a key role in understanding the world and its peoples. A forward-looking geography also requires what one might call *temporal* exploration; that is, mapping the possible nature of short- and longer-term futures. To turn our backs on the future is to avoid taking responsibility for the consequences of our choices and actions in the present.

In a recent issue of *Teaching Geography*, Wright (1998) argued for the importance of studying the 'near future' in geography, an endeavour which I would heartily endorse. The piece began by implying that such a task largely consists of making attempts to predict the future which then inevitably turn out to be wrong. However, this is not at all what the international field of futures studies is about. It is concerned, rather, with *how* people think about the future, how this affects their *choices* in the present and how they can act to bring their *preferred* futures about. This is a much more creative and worthwhile process and one in which geographers are already engaged (Masser *et al.* 1992).

Young people are, of course, very interested in the future. After all, they are going to spend more time there than their teachers will. Yet we do not often ask them what they think and feel about the future or consult them about their hopes and fears for the future. The study described in the last chapter provides evidence about students' views of the future. The main issues that concerned the secondary pupils locally and globally are shown in Table 4.1.

*Table 4.1* Hopes and fears for the local area (this is indicated %)

| Hopes for the local area (%) | | Fears for the local area (%) | |
|---|---|---|---|
| Better amenities | 45 | More crime | 59 |
| Less crime | 41 | Unemployment | 39 |
| Less pollution | 33 | More pollution | 37 |
| Greater prosperity | 33 | Worsening environment | 29 |
| Environmental concern | 32 | Urban spread | 25 |

As this table shows, young people's hopes and fears often mirror each other. From this evidence it is clear that their concerns fall into three broad categories: the environment, personal safety and quality of life. We might respond to these in our geography teaching by

- asking students to list some of the things that they like and dislike about their local area;
- asking students to identify individually three hopes they have for the future of their local area;
- then asking them to identify individually three fears they have about the future of their local area.

We can then ask ourselves:

- How many of these relate to geographical themes?
- How do they help us understand our students' relationship to their community?

Again, hopes and fears for the wider world mirror each other. Young people are clearly aware of the major global issues of environment, war and peace, poverty and

*Table 4.2* Hopes and fears for the global future

| Hopes for the global future (%) | | Fears for the global future (%) | |
|---|---|---|---|
| No war/more peace | 79 | More wars | 67 |
| Less pollution | 44 | More pollution | 45 |
| No poverty/hunger | 38 | More poverty/hunger | 28 |
| Good relationships between countries | 38 | Global warming | 25 |

hunger. We might respond to these hopes and fears in our geography teaching by asking students:

- What are some of the events going on in the wider world that they are aware of? How do they feel about these?
- To list individually three fears that they have in relation to the future of the world.
- Then to list individually three hopes that they have for the future of the world.
- How many of these relate to geographical themes?
- How does it help you to understand your student's perceptions of the world in which they live?

## Future geographies

While it is widely recognised that geography has a major contribution to make to the spatial dimension of the curriculum, it also has a significant contribution to make to the temporal dimension. Geography has, of course, always been concerned with time, in that it is interested in the nature of change, both natural and human. This is clearly set out in Eleanor Rawling's *Programmes of Study: Try This Approach*, which includes the following in a list of key geographical questions:

> *What might happen? With what impacts? What decisions will be made and with what consequences?* Predicting how the place or situation might change; analysing how decisions will be made; evaluating the impacts and consequences of these changes.
>
> *What do people think about this and why? What do I think and why?* Becoming aware of what a place or situation might be like in the future, and of how it may affect people; evaluating alternatives for the future; developing a personal response and justifying this.
>
> (Rawling 1992: 12)

We need to look, therefore, at how places have changed in the past and how they are changing now and how they may change in the future. We need also to look at how the environment has been changed in the past, how human activity is changing it now, and how it might be changed in the future. This will give us opportunities to address many of the concerns which students have about both their local and the global community. It means that we will be helping young people to think more critically and creatively about the future and therefore helping them to develop some of the key skills of responsible citizenship.

Many teachers, in all age phases and all subject areas of the curriculum, are interested in how they can educate their students more effectively for the future (Fountain 1995). What follows are specific suggestions as to how this interest can be activated within geography. They are drawn from the teachers' resource book *Citizenship for the Future: A Practical Classroom Guide* (Hicks 2001).

## Looking at change

Change is an ever-present theme in geography, and young people find this most interesting when it is brought out into the open and examined explicitly. Figure 4.1, for example, invites students to consider and record changes that they are aware of in their own lives, their local community, in Britain and the world.

Simple sketches and a keyword are put in each box. The resulting posters can create a really eye-catching display. This activity provides teachers with a snapshot of how students perceive their changing world. The matrix can also be used to explore change in one particular locality or in relation to a specific theme such as the environment. A variety of questions can be used to initiate wider discussion.

| Things that are changing | | | |
|---|---|---|---|
| In my life | Locally | In Britain | In the world |
|  |  |  |  |
|  |  |  |  |
|  |  |  |  |
|  |  |  |  |
|  |  |  |  |

*Figure 4.1* Things that are changing

## 44 Lessons for the future

- Which changes interest students most and why?
- Where do they get their information from?
- How accurate is it?
- What are the causes of these changes?
- Who benefits from these changes and who loses?
- Are they changes that students welcome or not?

Use the matrix shown in Figure 4.1 to find out what sorts of change your students are most aware of.

- What are their sources of information?
- What is it that interests them about these changes?

Devise a similar matrix to record recent and current changes in the local area – for example, in relation to roads and buildings, the environment, rivers and weather. Explore the differences between natural and human change.

Repeat this activity for a contrasting area in the United Kingdom.

- How similar or different are the changes that are occurring?
- What are some of the reasons for these changes?

Having accustomed students to observe change more closely, it is then possible to set this within the wider framework of past, present and future. One way of doing this is with the idea of the 'extended present'. In many communities today there will be some people who are 100 years old and some of the babies born today may live to be 100. In our own and other communities we can therefore stretch out and touch both past and future across a 200-year period – the extended present. This most readily comes alive when shown as the relationship between generations (see Figure 4.2).

Although not all students may know their parents or grandparents, the emphasis here is on the changing generations. Students could ask their parents:

- What stories can they each tell of this place?
- What photographs, artefacts or memories would they like to share?

Studying the past of a place in this way – how it became as it is today – is not unusual, but the extended present idea takes this further.

- What changes might occur in the future if present trends continue?
- When might students themselves become parents or grandparents?
- What responsibilities do we have towards future generations?

Change is not something that just happens to us. It is something to which we contribute, whether by our actions or lack of actions. If an out-of-town supermarket

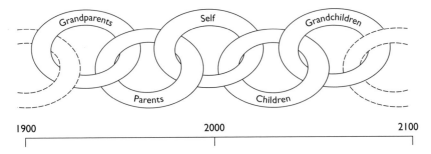

Figure 4.2 The extended present

is planned (with all the consequent impacts on local shops and traffic), we can support it, oppose it or do nothing. Students should be aware that doing nothing is, of course, always a vote for the most powerful group in any situation. Students could investigate a particular locality over the last 100 years and into the future. The same questions can be applied to different topics, as in the examples below.

### Farming

- How and why has the form of farming changed (crops, markets, pesticides, mechanisation)?
- How may farming change in the future and why?

### Transport

- How and why has transport changed its form (type, efficiency, routes, volume)?
- How may transport change in the future and why?

### Own locality

- How and why has your community changed its form (size, function, land-use, industries)?
- How may your community change in the future and why?

## Using timelines

Looking at change and the future does not necessarily have to cover a 200-year period. A simple timeline, drawn to map change in the present and the future, might cover a year, a decade or longer. The important point is that timelines can be used to explore future possibilities – that is, the *consequences* of change. One way of doing this is to look at current trends; for example, to do with land-use, employment or migration in a locality or in relation to an environmental issue. How might these trends affect the future of this place or issue? The possibilities can be sketched onto a timeline with dates and annotations.

# 46 Lessons for the future

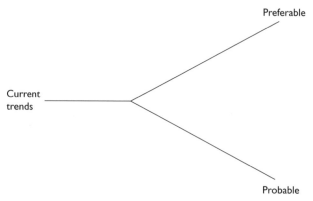

*Figure 4.3* Alternative futures

Drawing future timelines immediately raises a major question for students. They will want to know: is this the future that we think is *likely* to happen, or is it the one that we want to happen? This question highlights one of the most crucial distinctions in thinking about the future. On the one hand, we have *probable futures* – that is, all those which seem most likely to come about because of current trends. Probable futures focus on what people most expect to happen in a given location or in relation to an issue. By contrast, *preferable futures* are all those that we would most like to come about. Preferable futures reflect our deepest values, wishes and priorities. Probable and preferable timelines can be drawn separately or together, as shown in Figure 4.3.

Such a diagram could be used to map probable and preferable futures for your street, your town or indeed any other location. If doing both, it is best to complete the probable timeline first. Older students' probable futures sometimes reveal a measure of concern because they assume that little will change. As an example, my own probable future for the issue of traffic in Bath would show increases in traffic congestion, air pollution and respiratory diseases. My preferable future, by contrast, would show alternatives to reliance on the car, including a mix of buses, pedestrianised areas, tram routes and more cycle ways. This can lead students to the question: who else in my town is interested in this sort of future? Investigation then reveals that all sorts of organisations are concerned about traffic and the local environment. Students learn about positive change at grassroots level in their own and other communities. Finding out about, and being involved in, local action for change can go a long way in assuaging young people's concerns for the future.

- Take a particular environmental issue that the class has been studying and ask students in pairs to draw a probable timeline to show its future development. What changes do they predict and how might these affect people?

- For the same issue, ask students in pairs to draw their preferable timeline for its future development. How is this different? What needs to be done to bring such changes about?
- Ask students to draw probable and preferable timelines for a distant locality that they have been studying. Which future influences do they think are local to the area and which are global in affecting that place?

## Using scenarios

A further way of examining what might happen and with what consequences is through the use of scenarios. Scenarios are like maps, sketches or short stories about the future. Their purpose is to catch students' attention and their imagination by highlighting different aspects of possible futures. Thus, while any number of futures is actually possible for a given locality, it is useful to identify, say, four quite different scenarios. These can be used to explore how different decisions in the present will lead to different futures. The point of scenarios is to show a *range* of possible futures. One of my favourite examples was a leaflet, 'Landscapes for Tomorrow', from the Yorkshire Dales National Park (1989). This included a series of sketches showing different future landscapes for the National Park. Each had an accompanying paragraph that explained what decisions had been made in the present to bring that particular future landscape about.

If we take society more broadly, popular views of the future tend to fall into four broad categories, as shown in Figure 4.4 (see Hicks 2001 for more detail). Each view of the future is summed up by its title: 'More of the same'; 'Technological-fix'; 'Edge of disaster'; 'Sustainable development'. Students can 'visit' each of these futures to carry out 'fieldwork' investigations by asking the following questions:

- Do you think people like living in this possible future?
- What are some of the good things about it?
- What are some of the difficult things about it?
- Who will benefit and who will lose in this future?
- Say why you would or would not like to live in this future?

Such scenarios are much more than fanciful thinking. They are a widely used tool for assessing options and weighing up alternatives. Teachers may wish to draw their scenarios or let students devise their own. The purpose here is to encourage more forward-looking thinking, consideration of alternatives and consequences, so that wiser choices can be made in the present. It is about helping young people to develop skills of foresight for use in their own lives and as responsible citizens. As stated earlier, such aims are at the heart of good geography teaching: predicting how a place or situation might change, evaluating the impacts and consequences of changes, evaluating alternatives for the future, developing a personal response and justifying this.

# 48 Lessons for the future

Figure 4.4a More of the same

*Figure 4.4b* Technological fix

## 50 Lessons for the future

*Figure 4.4c* Edge of disaster

A geography for the future 51

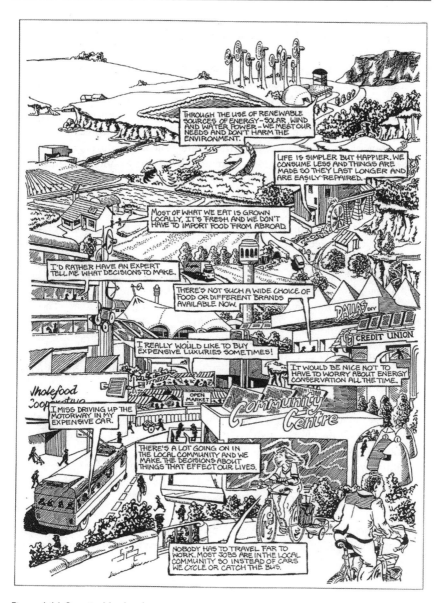

*Figure 4.4d* Sustainable development

- Students draw up alternative scenarios for the future of their classroom and/or the school grounds. (This needs to be a real exercise in which students have an actual say about the changes.)
- Alternative scenarios could be drawn up for the local shopping centre, a revised transport system or the local community as a whole. Students can then find out who is already doing this sort of work in the community.
- Students draw up a set of alternative scenarios for a contrasting locality; for instance, to show different possible impacts of tourism in an area like the Lake District, or the consequences of different aid projects to a developing country.
- Scenarios could be drawn to explore different energy options for the future in relation to, for example, energy efficiency and renewable sources or global climate change.

The 2002 Earth Summit (Dodds 2001), which will review progress over the ten years since Rio, is a reminder to teachers that issues of environment and development will be on the global agenda for much of the twenty-first century. The global agenda is also the local one and many exciting initiatives have taken place under Local Agenda 21 programmes. Young people need to explore the notion of sustainability, which is of little interest to them if delivered merely as an abstract concept (Huckle and Sterling 1996). They also need to be consulted about their visions for the future, on which they have many interesting things to say. Geography can make a major contribution to a more explicit futures dimension in the curriculum – an essential element in any effective education for the twenty-first century. You can make a start (if you haven't already done so) by:

- finding out about Local Agenda 21 initiatives in your area – in some places young people are already involved in the process of planning the future of their community;
- taking your class to visit the Centre for Alternative Technology near Machynlleth in Powys (see Centre for Alternative Technology 1995). There may be somewhere similar in your own region.

One of your tasks as a teacher is to help your students think more critically and creatively about the future. How have you helped them to clarify their visions for a more sustainable future?

# Chapter 5

# Towards tomorrow
## Strategies for envisioning the future

## Summary

One of the main concerns of social and environmental education is the need to create a more just and ecologically sustainable future, yet such educators often lack the expertise needed to explore alternative futures in this way. However, there is much of value that can be learned from futures studies and from the work of futurists. In particular, futures studies is interested in how people view the future and ways of cultivating clearer images of preferred futures. One established strategy for this is the futures workshop, and examples are given of work in this field by Robert Jungk, Warren Ziegler and Elise Boulding. It is suggested that a valuable cross-fertilisation of ideas could occur between educators and futurists.

Social and environmental educators spend a lot of time these days talking and writing about the need for a more just and ecologically sustainable future. Indeed, this has become the hallmark of much of the more radical debate about social change. However, in their debate about preferred futures, educators often still seem unaware of work by futurists on which they could profitably draw. The purpose of this chapter is to help bridge the gap between these fields of expertise.

## Preferred futures

Duane Elgin writes:

> Although there are numerous ecological, engineering, and energy requirements for a sustainable future, there are three more subjective requirements that seldom get the attention they need. These are: i) a simple and compelling story of the future; ii) practical ways to participate in creating that future in our daily lives; and iii) a conscious democracy that communicates its way into a sustainable future.
>
> (Elgin 1991: 77–8)

In discussing the first of these points, Elgin stresses that

> We cannot build a future we cannot imagine. A first requirement, then, is to create for ourselves a realistic, compelling, and engaging vision of the future that can be simply told. If our collective visualisation of the future is weak and fragmented, then our capacity to create a future together will be commensurately diminished.
>
> (1991: 78)

I have previously argued both that the school curriculum needs to become more future-orientated and that the future *per se* has been a missing dimension in fields such as environmental education. In particular, this chapter highlights the importance of futures studies and futures workshops to those working in education. Both of the above help to meet Elgin's three 'subjective requirements' for creating a more sustainable future.

## The value of futures studies

Environmental and social educators clearly have an ongoing interest in the future. Much of the concern over such issues in the last thirty years has arisen because people wish to avert a future in which environmental and social degradation are the norm. However, this interest in the future is generally implicit rather than explicit. Consequently, it often lacks rigour and much of the debate about such alternative futures loses its edge. This deficiency can easily be overcome by matching the expertise of such educators with that of futurists and others working in the futures field.

Beare and Slaughter (1993) argue that the futures field comprises three overlapping concerns: futures research, futures studies and futures movements. Futures research is concerned with economic and technical forecasting, strategic planning and technological change and is largely carried out by government departments and big corporations. 'Futures studies' is the term used to describe the work of academics, educators, critics and commentators who wish to communicate futures ideas to a wider audience. Futures movements are those new social movements which have a particular futures orientation and vision of a preferred world, whether in relation to peace, justice or the environment.

The World Futures Studies Federation and the World Future Society cater for the interests of futurists and others, as well as journals such as *Futures* and *Futures Research Quarterly*. Some countries have set up institutes of foresight (Slaughter 1995). Writers such as Slaughter (1996; 2000) and Bell (1997) have helped map out the futures field, while others, such as Hazel Henderson (1993) and Graham May (1996), explore futures-related issues in other rich and insightful ways. Clearly, many authorities, although not futurists *per se* (such as Brown *et al.* 2001b), also have a particular interest in the future.

Jim Dator, director of the Hawaii Research Center for Futures Studies, writes:

> 'The future' cannot be 'studied' because 'the future' does not exist. Futures studies does not – or should not – pretend to study the future. It studies ideas about the future (what I usually call 'images of the future') which each individual and group has (often holding several conflicting images at one time). These images often serve as the basis for actions in the present. Individual and group images of the future are often highly volatile and change according to changing events or perceptions. . . . Different groups often have very different images of the future. Men's images may differ from women's. Western images may differ from non-Western images and so on.
>
> (Dator 1996a: xix–xx)

He also emphasises a further aspect of futures studies; namely, its particular interest in identifying preferable futures.

> The future cannot be 'predicted', but 'preferred futures' can and should be envisioned, invented, implemented, continuously evaluated, revised and re-envisioned. Thus, another major task of futures studies is to facilitate individuals and groups in formulating, implementing and re-envisioning their preferred futures.
>
> (1996a: xx)

Now clearly, these are the sorts of concerns that are of interest to many educators. What sort of images of the probable future do people have at the beginning of the twenty-first century? And are their preferable futures ones that highlight social and environmental concerns?

The research on images of the future over the last thirty years has been somewhat sporadic (see Chapter 3). One of the largest studies, carried out by Ornauer *et al.* (1976), compared views of the future in ten different countries. The overall picture was not an encouraging one, and people's ability to think about and visualise the future was found to be not very well developed. By and large, pessimistic visions of the future were more common than optimistic ones. Galtung (1976) comments that most societies have an elite whose job it is to think about the future, so that people do not expect to be given any say or to be consulted about the sort of social future they might want.

It is interesting to recall at this point the thesis of Polak (1972) who, in his study of images of the future in Western society, argued that such images act as a mirror of the times. By this he meant not just that people's concerns necessarily reflect the times they live in, but that the pessimism or optimism of such images has much to say about the health and well-being of that society. He suggests that there is a close correlation between the rise and fall of images of the future and the rise and fall of society and culture itself. As long as a society's images of the future are positive and flourishing, they act like a magnet drawing society on towards its envisioned future. Once such images begin to decay and lose their vitality, however, then culture cannot long survive. Writing during the 1950s, Polak saw the mid-twentieth century as being unique in only possessing negative images of the future. Nearly fifty years later, at the end of the twentieth century, positive images of the future are still hard to come by.

## Futures workshops

As a result of Polak's work several futurists became interested in working with images of the future and developed a range of strategies for doing this. The work of three such futurists is described here as being of particular value to environmental educators. One of the earliest proponents of futures workshops was Robert Jungk, a major European futurist who, with Johan Galtung, helped found the World Futures Studies Federation. In the introduction to *Future Workshops: How to Create Desirable Futures*, Jungk writes:

> I should first explain how I came to be interested in starting this democratic institution, the future workshop. I was a victim of Hitler's regime, leaving Germany in '33. I felt powerless about the holocaust. . . . Ever since then I have looked for ways that people can fight back and can influence the course of events. The future workshop is just such a way. It helps people to develop creative ideas and projects for a better society. For trying to resist something is just part of the story. It is essential for people to know what they are fighting for, not just what they are fighting against.
> (Jungk and Mullert 1987: 5)

From the early 1960s onwards Jungk ran workshops all over Europe with a wide range of community, business, government and activist groups. Such workshops last for a day, a weekend, or longer, and have four main phases. In the preparatory phase, participants state why they have come to the workshop and its structure is explained to them. In the critique phase, complaints and criticisms are collected in order to identify the key components of the problem. This is followed by the fantasy phase, in which various processes, such as brainstorming, are used to generate a series of 'utopian schemes' that might resolve the problem. In the implementation phase, the most popular suggestions for action are identified and checked for practicability. At a continuing workshop, participants review their choice of project and finalise detailed action plans.

Described in this way the process sounds a little dry, but it comes alive in the comments that Jungk makes about his experiences. At first he did not realise how difficult it would be to get people to think creatively about the future. Sometimes groups were hostile because they expected him to provide them with the answers. Often people were passive, he realised, because of 'the hostility of our social environment to anything from the realm of the imagination' (Jungk and Mullert 1987: 17).

> In our discussion we relived the shattering of children's worlds and the entombment of their imagination. We found, in so doing, we were able to dig it out, somewhat the worse for wear but still alive amid the debris of all the derided and trampled dreams.
>
> (1987: 17)

In particular, Jungk stresses the need to liberate the intuitive and emotional in these workshops as well as using the rational and analytic. The most constructive solutions, he found, often came from clumsily expressed yearnings and dreams rather than the use of 'scientific' logic. Whether it was a mining community due for resettlement, a group concerned about the social impact of computer technology or problems of urban living, Jungk found that people's creativity could be tapped to envision their preferred futures and then to act on them.

In the United States, Warren Ziegler has worked since the early 1970s to popularise the notion of futures workshops. Like Jungk, he has worked with a wide range of groups, from business executives and urban planners to hospital trustees and city councils. What characterises Ziegler's work, however, is a much wider repertoire of processes for envisioning preferred futures. He writes:

> I much prefer to call futures-invention an art, a craft, a discipline, a set of practices. Put all together, these become an approach to the future which we adopt in order to bring about a future different from what it might otherwise be were we not to intervene, deliberately and creatively, in the present to 'invent' it.
>
> (Ziegler 1987: 1)

In particular, he stresses that the most important thing about the future is not what we know (that is, forecasting) but what we intend, for from intention comes action. Workshop participants are therefore invited to discover their intentions towards the future by engaging in the practice of imaging. Such imaging or guided visualisation can be used to identify the details of a group's preferred future. 'The future . . . is an act of the imagination. Our aim . . . is to discern our intentions towards the future so as to shed light on our present situation and provide guidelines for changing our actions so as to move towards that intended future' (Ziegler 1987: 8).

A detailed exposition of Ziegler's work is to be found in *Envisioning the Future: A Mindbook of Exercises for Futures-Inventors* (1989). This manual contains a variety of exercises all of which stretch and extend the participant's imaging skills. Detailed scenarios are built up, at first individually and then in groups, to create shared visions of the preferred future. Careful attention is paid to how such scenarios connect to the present and the action needed to bring them about. There are five stages in Ziegler's (1991) futures workshops: (1) the discerning of concerns; (2) focused imaging; (3) creating shared vision; (4) connecting the future with the present; (5) strategy paths and action. These stages are similar to the cycle used by Jungk but the specific exercises are more elaborate and sophisticated.

Another major figure in the field of envisioning futures is Elise Boulding. Her initial interest came from translating Polak's work (Boulding 1979). This led her to collaborate with Ziegler on a joint venture in the early 1980s. In working with peace activists Boulding came to realise that, although they were clear about what they opposed, they found it difficult to visualise what they were for. In short, such activists found it difficult to imagine what a world without nuclear weapons might actually look like. The outcome of this collaboration with Ziegler is described in *Building a Global Civic Culture* (1988), in which Boulding also more broadly explores the role of the social imagination in global citizenship.

Boulding is one of the few people to have detailed the outcomes of futures workshops and to have analysed the futures imagery that arises. In particular, she notes the presence of several common themes.

> Whatever people are doing, women and men are doing it together. Children and the elderly seem to be everywhere – there is no age segregation. Communities are also described as racially mixed. Learning seems to take place 'on the job'. This is a non-hierarchical world; no one is 'in charge'. It is also one in which locality is very important. . . . Technology is low profile, but it does exist and everyone reports that it is shared. . . . Although the new-consciousness theme varies in importance, there is widespread reporting that people are operating out of a different sense of awareness than that of [the present].
> (Boulding 1994: 73)

Boulding describes this as a 'baseline future', common to most imaging groups that she has worked with. Two other images she reports, although not universal ones, are a sense of boundarylessness (a greater flow of people and ideas unimpeded by formal structures) and the use of the adjectives 'bright', 'clean', 'green' to describe this preferred future world. She suggests that the recurrence of such images, regardless of the participant's background, might indicate some 'deep structures at work in the futures-imaging process that should be more fully explored' (1994: 82).

These three examples of futurists at work highlight the richness of such processes as tools for social change. They have also been used extensively by Dator (1993), for example, in relation to state judiciaries and by other facilitators working with organisations and communities (Weisbord and Janoff 1995). It is clear that the social imagination has a crucial role to play in identifying the futures that we wish to create. What is also certain is that Western society undervalues non-cognitive ways of knowing and thereby marginalises the very tools needed to envision our way out of current situations.

However, as Meadows points out:

> We should say immediately for the sake of skeptics that we do not believe it is possible for the world to envision its way to a sustainable future. Vision without action is useless. But action without vision does not know where to go or why to go there. Vision is absolutely necessary to guide and motivate action. More

than that, vision, when widely shared and firmly kept in sight, brings into being new systems.

(Meadows *et al.* 1992: 224)

Environmental and social educators have a specific concern that the future should be different from today. In order to elaborate critically on the sustainable futures that they might desire they can draw on the insights both of futures studies and of futures workshops. In so doing they can help provide Elgin's (1991) 'compelling story of the future' and provide practical ways to participate in creating that future. That environmental educators have now begun this process is shown in Rickinson's (2001: 249–51) critical review of environmental education research. Such a cross-fertilisation of ideas will undoubtedly prove to be most fruitful.

Chapter 6

# Retrieving the dream
## How students envision their preferable futures

### Summary

Popular images of the future have a crucial role to play in societal development, but there is still a general dearth of positive images in the Western world as we begin the twenty-first century. In this context futures workshops are particularly valuable as a process for enabling participants to envision the future more clearly. Attention is focused on the work of Elise Boulding, and a study described which utilised her procedures in order to help students identify the key features of their preferable futures. Some support is found for Boulding's contention that a 'baseline' future often seems to emerge from such work.

### A sense of the future

> As the year 2000 approaches, releasing a rush of millennial hopes and fears, I take for granted that the future will once again play a dominant role in our lives ... a few romantics like myself still believe that our sense of the future remains intact, a submerged realm of hopes and dreams that lies below the surface of our minds, ready to wake again as one millennium closes and the next begins.
> (Ballard 1994: 1)

In his introduction to *Myths of the Near Future*, novelist J.G. Ballard (1994: 1) argues that at some point in the 1960s our 'sense of the future seemed to atrophy and die', and he looked to the millennium for a reawakening of interest in the future. Maybe one of the tasks of futurists and futures educators has been to keep that interest alive. If this is the case, it would seem that they have not been terribly successful. One major study (Ornauer *et al.* 1976), for example, suggested that people were not very accustomed to thinking about the future and, when they did, they framed this largely in terms of developments in science and technology. Livingstone's (1983) study identified a widening of concern to other social issues but still found that the best imagers of the future were corporate capitalists.

Many authorities – such as Bell and Mau (1971), Huber (1978), Wagar (1992) and Boulding (1988) – have argued that images of the future have a crucial role to play in societal development and that study of such images should thus be a central concern of futurists. Dator (1996b) argues that one of the things futures studies does is to help people clarify their hopes and fears about the future in order to move beyond passive forecasting to the generation of preferred futures as a basis for planning and action. Perhaps the most provocative proposal comes from Polak's (1972) study of changing images of the future over the last 3,000 years in Western society. In examining the rise and fall of such images he argues that, as long as a society's image of the future is positive, its culture will blossom but that, once the image begins to decay, the culture will lose its vitality and decline.

At the very least this suggestion should make us look more closely at popular images of the future today and during the twentieth century. Clarke (1992) argues that the high expectations held for the future at the beginning of this century were drastically dashed by two world wars. After a short bout of technological optimism in the 1950s, the full horror of the nuclear arms race again made the future a place of fear for another thirty years.

Public images of the future, as Wark (1996) points out, are often largely negative. Popular movies from *Mad Max* and *Bladerunner* to *Judge Dredd* and *Waterworld* provide a regular diet of dystopia. As Lowenthal (1995: 385) notes, 'Disaffection with the future stems in part from its growing uncertainty. The more complex we see our world to be, the harder it becomes to predict outcomes of our actions.' Research into young people's views of the future suggests that they are well aware of the turbulence of these times and the hazards that this may bring (see Chapter 3).

In the face of this apparent dearth of positive images one is forcibly drawn to the question of preferable futures. Polak (1972) remarked, in the 1950s, that there were few, if any, positive images of the future to be found in Western society. At first glance the situation would seem to be little better at the start of the new century. It should be a crucial concern for all educators if young people often have negative images of the future. Motivation in the present is greatly affected by images of personal and social futures. If, on the broader level, Polak's thesis is correct, then this state of affairs becomes even more worrying. How might such a situation begin to be resolved?

## Envisaging the future

Perhaps there *are* signs of a reawakening of interest in the future. Faced with the social, technological and environmental dilemmas of our times, various authorities are urging a reappraisal of procedures for envisioning alternative futures. Typical of this call is that by Meadows, *et al.* in *Beyond the Limits: Global Collapse or a Sustainable Future*.

> Visioning means imagining, at first generally and then with increasing specificity, what you really want. That is, *what you really want*, not what someone has taught you to want, and not what you have learned to be willing to settle for. . . . Some people, especially young people, engage in visioning

with enthusiasm and ease. Some people find the exercise of visioning painful, because a glowing picture of what *could be* makes what *is* all the more intolerable.

(Meadows *et al.* 1992: 224)

Those futurists who work at the normative end of the futures spectrum have, of course, long explored processes and procedures for envisioning the future, and some of the key work in this field has been described in Chapter 5. Envisioning has also come to the fore as a tool used in creative problem solving, for developing leadership skills, in community visioning and for building shared vision in organisations (Markley 1992; Nanus 1992).

The incentive for the research described in this chapter came from the work of Boulding and Ziegler (see Chapter 5). It was also prompted by Rogers and Tough's (1992) account of the dilemmas raised when teaching students about futures issues. This echoed my own experience, which is that many students cut off when they realise the extent and nature of the global crisis. The discovery that they share similarly pessimistic probable futures can encourage despair rather than solidarity. The crucial pedagogical question then becomes how to provide stories of hope which both liberate and empower. One method, of course, is to explore projects which clearly embody visions of a more sustainable future (see Chapter 9). Another is to involve students in the envisioning of their *own* preferable futures.

Evaluative accounts of futures workshops are not easy to come by, but Elise Boulding provides a rich source of information in describing her experiences in running such groups. She describes the methodology of her workshops as being based on

Polak's concept of a 'breach in time', a drastic discontinuity between present and future that can nevertheless be encompassed by the human imagination. Participants must step, in fantasy, into a future very different from the present, and report back from that future on their observations of a society, which they then must analyse in terms of the social institutions that could sustain it.

(Boulding 1994: 67)

Details of the procedures used by Boulding can be found in *Preparing for the Future* (Hicks 1994b) and an outline of that process is given later. Boulding describes the results of working with three specific groups in the United States: the first on the changing role of men, the second an international youth conference on peace, the third a group of scholars working on models of a warless world.

The purpose of these three workshops was to see whether people could image a hopeful future given an appropriately facilitative environment, and whether any major differences would arise between the groups. Each workshop lasted for three hours and some 300 participants were involved in all. They began by individually constructing a wish list for the future. What would they like to see in the best possible future they could imagine? This was followed by a guided visualisation, in which they 'stepped into' their preferred future as if it had already happened. People were encouraged to identify exactly what had changed and to look for the proof that demonstrated that their preferred future had actually come about. After this visualisation, they worked in small groups to create shared scenarios recorded in the form of posters.

Boulding analysed the images, key words and concepts in the posters, and counted their frequency of appearance, and themes were then combined into categories. It was clear that the process *did* enable participants to envision a hopeful or preferred future, so attention then turned to the types of imagery generated in the workshop. Some of the common images arising in her workshops are shown in Table 6.1 below.

*Table 6.1* Elise Boulding's 'baseline' future

- a lack of divisions based on age, race or gender
- a non-hierarchical world with no one 'in charge'
- a strong sense of place and community
- low-profile and widely shared technology, particularly relating to communications and transport
- people acting out of a more peaceable 'new consciousness'
- education generally taking place 'on the job'
- food grown locally
- a 'bright, clean, green' world (particularly strong among young people)
- a sense of the local community as joyful, nurturing and celebratory
- a 'boundaryless' world unimpeded by social, occupational or political barriers

Boulding (1994) reports that these features are common not only to the three particular groups she describes, but to most imaging groups that she has worked with. These features, she argues, may be thought of as a 'baseline future'. Other vivid images related to the 'feel' of the local community, a sense of nurturing support, joyfulness and celebration. These descriptors provided a rich resource against which to compare this UK study. Would participants come up with similar images in relation to their preferred futures or something completely different?

## Working with students

Having been fascinated by Boulding's account of carrying out such workshops, I wondered both how replicable they were and whether they might go some way to counteracting the despair that students often feel when studying global issues. Boulding's workshops also had an action planning component to help bridge the gap between image and action, but this often had to be dealt with in follow-up sessions due to the constraints of time. I decided, first, to follow Boulding's own procedure as closely as possible in order to facilitate a comparison between findings. I had three questions in mind:

- Does this process help students generate images of preferred futures?
- What is the range of futures imagery that consequently arises?
- How do participants respond to such a workshop process?

Workshops were run with an opportunity sample of students studying at three institutions of higher education in the south-west of England. Most, but not all, of the ninety students involved were training to be teachers. Most were in their late teens or early twenties but there was also a sprinkling of mature students. The three groups were, respectively, post-graduates training to be teachers, students studying a module on global issues as part of a BA/BSc combined award and BEd students training to be teachers. Ideally, I wanted a half day for each workshop, for it felt as if the process demanded at least this length of time in order to be effective. However, it was not possible to get more than three hours with any of the groups, the same length of time as Boulding had.

Each workshop was introduced as an opportunity for participants to reflect on the future and, in particular, to clarify their preferred world in 2020. The context they were given was Bristol (or wherever they lived) in the world, the world in Bristol. They were thus encouraged to consider both local and global possibilities. Few of the students had specifically engaged in futures thinking before and most had not met me before. After a brief introduction to the purpose of the workshop, participants began by drawing a global timeline, working in pairs. This line began in the 1990s and then branched to give a probable future and a preferable future for the period 1995–2020. On the first line they marked events, trends and issues that they *expected* to come about. As with most of the student or teacher timelines I have ever seen, these were generally pessimistic in tone. On the second line they indicated the main events, trends and issues that they *hoped* would come about in that period. In neither case were they given any directive as to what they should include. This arose solely out of the paired discussion.

For the next stage of the workshop they were asked to focus on the key elements of their preferred future, those things which they most wished to see by the year 2020. The second activity involved use of the creative imagination, using a guided visualisation designed by Boulding to 'visit' their preferred future. Students were reminded that not everyone sees images when using the imagination and that feelings or intuition may speak more clearly for some. It was emphasised that this was not an exercise in prediction but an opportunity to 'see' their desired future. Short periods of silence were interspersed with brief instructions. In particular, participants were asked to look around them for proof that their preferred future had actually come about. They were asked to note the ways in which people and society were different and to look for evidence that their desired changes had come about.

The third activity involved participants working in the same pairs to share their futures with each other. It was emphasised that this was a time for non-judgemental listening. Each pair was asked to create a poster illustrating the main features of their preferred world in 2020. They were asked to draw this in a way that made it self-explanatory to others. This resulted in a fascinating set of posters that were then

*Table 6.2* Students' preferred futures 2020

|  | % |
|---|---|
| *Green* – clean air and water, trees, wildlife, flowers | 79 |
| *Convivial* – cooperative, relaxed, happy, caring, laughter | 74 |
| *Transport* – no cars, no pollution, public transport, bikes | 55 |
| *Peaceful* – absence of violent conflict, security, global harmony | 53 |
| *Equity* – no poverty, fair shares for all, no hunger | 38 |
| *Justice* – equal rights of people and planet, no discrimination | 36 |
| *Community* – local, small, friendly, simpler, sense of community | 36 |
| *Education* – for all, ongoing for life, holistic, community | 30 |
| *Energy* – lower consumption, renewable and clean sources | 26 |
| *Work* – for all, satisfying, shared, shorter hours | 23 |
| *Healthy* – better health care, alternative, longer life | 19 |
| *Food* – organic farming, locally grown, balanced diet | 15 |

displayed for perusal by the rest of the group. The ninety students involved in this exercise produced forty-two posters for subsequent analysis. Time was given for group discussion about the visualisation and creation of the posters, which generally resulted in an animated debate. Finally, an evaluation sheet was completed by all participants.

After the three workshops the posters were analysed, as Boulding had done, to identify the main images that they contained. Each image was listed and then clustered by focus, leading to identification of twelve key themes overall. These themes, together with their associated images, are listed in rank order and summarised in Table 6.2.

Although this is a fairly simple form of analysis it nevertheless highlights the concerns of these students and the relative importance of different issues to them. Primarily, they want a future characterised by environmental concern, with nearly 80 per cent citing respect and reverence for the biosphere as their highest priority. The posters particularly referred to clean air, land and water, together with a richness and diversity of species, and an abundance of trees and flowers. Second, three-quarters of the students want a more convivial future; that is, one which stresses quality of life and human interaction. Reference was made particularly to a sense of social well-being, to a cooperative atmosphere, to less stress and more shared laughter. An awareness and celebration of our interconnectedness with the planet and with each other are thus the primary features of their preferred futures.

The next two key characteristics, put forward by just over half of the students, relate to conflict and transportation. Participants want to live in a future which is more peaceful, both locally and in the world as a whole. In one sense this is an extension of the conviviality that they wish to experience in their own lives to the wider community. It also involves a greater sense of security and harmony as a result of living in a less violent society. Transportation is the main issue that concerns this group, and in their preferred future cars have been banned and replaced by cheap and efficient public transport, particularly trams and cycles.

The baseline future for this group of students is therefore one which is green, convivial and peaceful, with a stress on alternative forms of transportation. Three other significant features are mentioned by just over a third of the participants. They want to live in a society characterised by much greater economic equity than today, referring in particular to an absence of rich and poor, food and homes for all, and fair trading leading to an end to Third World exploitation. Closely allied to this is a wish to see greater justice for all, an end to discrimination, respect for human rights and rights for all species. Also important at this level is the need to feel part of a community, combined with a sense that small is beautiful, whether in respect to settlements, business or agriculture. Education is mentioned by just under a third of the students as something which should be lifelong, available to all, community based and more holistic in approach.

The final four features of their preferred futures were mentioned by only a quarter or fewer of the students. Reference was made to use of alternative sources of energy; namely, solar, water and wind power; work being more congenial, satisfying and for all; more attention being paid to health care, including alternative approaches to health; and food being grown organically and locally.

In the evaluation, students were asked to indicate, on a five-point scale, the extent to which they found the workshop process easy to work with; whether it helped them to think more clearly about the future; whether it helped them to clarify their vision of the future; and whether it helped them to clarify appropriate action for change. As Table 6.3 shows, participants considered the workshop process to be highly effective in helping them both to think about and envision the future. It was less effective, however, in helping them to clarify relevant action towards bringing their preferred futures about.

What participants generally enjoyed most about the workshop was the guided visualisation and the chance to share with each other the details of their preferred futures. They were particularly impressed by the fact that they often shared the same hopes and fears for the future. One student commented, 'No one has ever asked me before what I really think about the future!' – a sad reflection on the futureless education that many young people still received at the end of the twentieth century.

*Table 6.3* Evaluation of workshop effectiveness

|  | % |
| --- | --- |
| Workshop process | 88 |
| Thinking about the future | 84 |
| Visions of the future | 77 |
| Action for the future | 50 |

(N=90)

## Some conclusions

Although one should be wary of generalising from this study, it does nevertheless add to the knowledge that we have about the value of futures workshops. While Boulding's groups were more varied in make-up than those described here, there is an interesting overlap in the futures described. Of the ten most common themes that she reports (Table 6.1), seven emerge in this study. These are: absence of discrimination; a sense of place/community; food grown locally; technology in relation to transport; peaceable; green; celebratory. This would go some way to support Boulding's contention that there is a 'baseline future' which often emerges from such workshops, a growing consensus on what the future we need should be like. Presumably, most of her participants were self-selecting in that they all had some concern about creating a better world. However, this does not invalidate the images that arise from the visualisation.

It is clear from this study that the process developed by Boulding does help students to generate images of their preferred futures despite the three-hour time constraint imposed on the workshops. It has also been possible to gain some idea of the range of images produced when using such processes. Participants generally responded with enthusiasm to the workshop process, which, according to their evaluation comments, was helpful in enabling them to think about and envision the future more clearly. Lack of time meant that less attention was paid to the possibilities of action planning, however, and this is reflected in the lower score for 'action' in Table 6.3.

What this study was not able to do was to estimate the effectiveness of such workshops in creating what Boulding calls 'action readiness'. Other reports, however, indicate that this particular process is effective in bringing about statistically significant change both in clarifying future goals and in creating a sense of empowerment (Spradlin *et al.* 1989). Certainly, futures workshops in general have made a significant contribution to citizen action and the concerns of a wide range of interest groups in society (Weisbord and Janoff 1995).

At the start of the new century it *is* possible to detect a quickening of interest in the future as Ballard predicted. However, this interest combines both hope and fear, for, as Kumar (1995b) argues, a sense of endings without new beginnings has been one of the constants through the twentieth century. It is little wonder that there were few positive images of the future as the millennium approached. Hobsbawm (1994), in his recent 'short history' of the century, argues that it has been the most murderous for which we have records. Quite clearly, guiding images of the future have a crucial role to play in giving society a positive sense of direction. It is as if we need to be given permission to reclaim our envisioning faculties at the end of such a dark century. We need to use this time as an opportunity to reflect on new beginnings and ones which, in particular, stress the need for a more just and ecologically sustainable society. We need, says Kumar (1995b), a sense of expectation *and* hope to take us forward. What these workshops show is that it *is* possible to retrieve the dream of a more equitable and humane world. If it can be imagined then surely it can be created too.

Chapter 7

# Stories of hope
## A response to the psychology of despair

### Summary

The new century, like the old, faces all the hazards and complexities of postmodern life which pose particular problems for those teaching about social, environmental and global issues. How does one stay close to the discomfort of such issues without falling into despair? In periods of rapid and turbulent change it seems crucially important to be able to identify the sources of hope and inspiration that educators and others may draw on. In an investigation designed to identify such sources the process of sharing came to be seen as a vital source of hope in its own right.

### Turbulent times

Theologian Thomas Berry, in discussing the dilemmas of contemporary life, argues that

> It's all a question of story. We are in trouble just now because we do not have a good story. We are in between stories. The old story, the account of how the world came to be and how we fit into it, is no longer effective. Yet we have not learned the new story.
>
> (cited in Milbrath 1989: 115)

Many writers have postulated that the Western world may be undergoing a cultural paradigm shift comparable to that of the Agricultural Revolution in the late Neolithic period or the Scientific Revolution of the eighteenth century (Capra 1983; Milbrath 1989; Inglehart 1997). Some have seen in this transformation the possibility of creating a more just and ecologically sustainable society (Meadows *et al.* 1992; Henderson 1993; Trainer 1995). The times are confusing, argues theologian Thomas Berry, because the old metanarrative of modernity has lost its validity and the new story has yet to be fully understood.

It is within this wider context that I wish to situate a research project that I am engaged in. Global and futures educators – that is, those who teach about global issues (whether the environment, wealth and poverty, human rights or peace and conflict) – are very much on the cultural front line in having to interpret the current state of the world to students (Steiner 1996). They are on the front line in the sense of choosing to teach about issues which threaten both our physical and emotional well-being. In the face of this there is always the danger of succumbing to the pessimism of the times (Bailey 1988). This is one reason why Huckle argues that 'optimism' should be a key component in environmental education:

> If we are not to overwhelm pupils with the world's problems, we should teach in a spirit of optimism. We should build environmental success stories into our curriculum and develop awareness of sources of hope in a world where new and appropriate technologies now offer liberation for all.
> (Huckle 1990: 159)

It is likely, however, that problem-orientated teaching still provides the main focus for much environmental and global education rather than success stories, although the importance of the latter has also been stressed by Bardwell (1991).

While the extent of the environmental crisis has been well documented by writers such as Merchant (1992), Seager (1993) and Redclift and Benton (1994), this needs to be placed within the wider context of global change (McGrew 1992; Brown *et al.* 2001). It is the contradictory forces of capital accumulation and globalisation that have helped birth postmodernity – or late modernity, as Giddens (1990) prefers to call it. Thus we find ourselves today contemplating what has been gained from the fruits of modernity: the Scientific Revolution, the Enlightenment, the Industrial Revolution, which together resulted in the great leap forward that enabled Western capitalist nations to dominate the affairs of the globe for several centuries.

The price of such 'progress' has, however, been high, since modernism valorised positivism, rationalistic planning, technocratic solutions, and the notion of inexorable progress, which in turn contributed to the nightmares and excesses of this century. Thus Harvey's reflection:

> Whether or not the Enlightenment project was doomed from the start to plunge us into a Kafkaesque world, whether or not it was bound to lead to Auschwitz and Hiroshima, and whether it has any power left to inform and inspire contemporary thought and action, are crucial questions.
> (Harvey 1989: 98)

Under postmodernity, social and cultural life is characterised by doubt, fragmentation and rejection of metanarrative. For some commentators this is cause for celebration: there is no longer any metanarrative that can be trusted, there are no 'big stories' that can help us make sense of the world. By contrast, we have a plurality of stories and any can have its day. The condition of postmodernity is thus one of acute uncertainty heightened by what Giddens (1991) calls high-consequence risks; that is, those over which the ordinary person has no control, such a global warming.

Humans respond to difficult situations in a variety of ways. It is a commonplace of psychotherapy that we repress material which we do not want to be consciously aware of, whether childhood or adult trauma. The process of therapy enables the client to re-engage with these split-off experiences and to integrate them healthily into everyday life, such that old wounds are gradually healed (Heron 1990). From that healing comes a sense of empowerment and an ability to face new situations.

Such repression can also be seen at work on a societal scale (Staub and Green 1992). Thus Lifton (1992) has written about the state of psychic numbing or dissociation which arises in response to particularly painful situations. To be willingly involved in the Nazi death camps or promotion of nuclear deterrence, he argues, requires that thought be separated from feeling. Indeed, to even contemplate the Holocaust or the effects of nuclear war may result in such shutting-off from feeling.

Postel recognises this when she writes:

> Psychology as much as science will . . . determine the planet's fate, because action depends on overcoming denial, among the most paralysing of human responses. While it affects most of us to varying degrees, denial often runs particularly deep among those with heavy stakes in the status quo. . . . This kind of denial can be as dangerous to society and the natural environment as an alcoholic's denial is to his or her own health and family . . . rather than face the truth, denial's victims choose slow suicide.
>
> (Postel 1992: 4)

Denial is an understandable response to the enormity of environmental and global issues. Indeed, a variety of responses are reported from students, ranging from anger and frustration to a sense of challenge and excitement (Rogers and Tough 1992; Hicks and Bord 1994; Hutchinson 1996). However, the emotional impact of global issues on students' learning is still a neglected area of research.

## The question of hope

In the light of the above it is vital to understand the sources of hope and inspiration that people draw on in postmodern times. Some notion of hope and optimism about the human condition necessarily underlies fields as diverse as sustainability (Meadows *et al.* 1992), utopian studies (Kumar 1991), community lifestyles (McLaughlin and Davidson 1985), radicalism (Button 1995) and futures education (Hicks and Slaughter 1998). Few writers have specifically explored the phenomenon of hope in detail, although two exceptions are Bloch's *The Principle of Hope* (1986), which brings together a mass of material from the Marxist and Romantic traditions, and Moltmann's *The Theology of Hope* (1967), which is an extensive exploration of hope in religious and secular traditions.

What both these authors do, from quite different perspectives, is remind the reader that concern for something Other, something better, something not-yet, is an inherent element in the human condition and one of the deep components of human creativity. It is hope that allows us to go on when conditions look bad or even impossible. It is hope that keeps possibility open. Thus Moltmann argues:

> Hope alone is to be called 'realistic', because it alone takes seriously the possibilities with which all reality is fraught. It does not take things as they happen to stand or lie, but as progressing, moving things with possibilities of change. . . . Thus hopes and anticipations of the future are not a transfiguring glow superimposed upon a darkened existence, but are realistic ways of perceiving the scope of our real possibilities, and as such they set everything in motion and keep it in a state of change. Hope and the kind of thinking that goes with it consequently cannot submit to the reproach of being utopian, for they do not strive after things that have 'no place', but after things that have 'no place *as yet*' but can acquire one.
>
> (Moltmann 1967: 25)

Several writers have recently acknowledged the importance of hope, though more often in implicit rather than explicit ways – for example, the essays in *The Right to Hope: Global Problems, Global Visions* (Thick 1996) and McKibben's case studies in *Hope, Human and Wild* (1995).

## Sources of hope

It was in the light of the above that an investigation was set up to explore the experiences of global educators who manage to retain a sense of hope and optimism while yet having to stay present to global threats on a daily basis. The principal research question was this: what are the main sources of hope and inspiration that such global educators draw on? The project was thus not attempting to measure the construct of hope nor to analyse its definition in the literature but to identify possible *sources* of hope.

In contemplating the framework for this research it was the researcher's personal experience of teaching global issues which helped define this research question. As Reinharz (1992) argues, such questions often arise from the blending of an intellectual question and a personal concern. The project was seen as a qualitative study within the interpretive paradigm (Cantrell 1993), designed to elicit some of the sources of hope that people can identify. It was decided, therefore, to work with a small, potentially information-rich, sample of global educators. A number of people were approached, all of whom were known to have an interest in global issues, and, based on availability, the final group comprised six men and six women. Some were LEA inspectors, others were responsible for educational projects or worked for NGOs.

Data triangulation was achieved through use of taped interviews, autobiographical writing and a focus group weekend. Participants first took part in a semi-structured interview of about an hour's length which allowed researcher and respondent to get to know each other and begin the process of reflection (Hitchcock and Hughes 1989). Questions were designed to elicit details of the interviewee's work and how they felt about global issues, and an initial exploration was made of possible sources of hope. The autobiographical writing was intended to evoke a more reflective response, looking at life events and people who had played a significant role in sustaining hope in difficult times. Finally, participants met for a residential focus group weekend which allowed them to share their stories and experiences. This was also valuable for participants, in that it provided an excellent forum in which to share and test out their views on others (Burgess *et al*. 1988). Each of these steps in the triangulation process achieved a gradual deepening of insight on the part of participants.

What follows is an initial account of what happened. It highlights what appear to be key insights for participants in their individual and joint exploration of their sources of hope. As with all interpretive research, other levels of inquiry begin to open up as these first impressions are described.

In their interviews and at the focus group weekend, participants were asked to identify those aspects of life in postmodern times which caused them most concern. At the global scale these included the continuing degradation of the planet, abuse of human rights, nuclear accidents, the power of multinational corporations, and hunger and poverty. Within the United Kingdom the group cited racism, Thatcherism, a decline in moral values, poverty and the state of education. Professionally, they felt stressed and overworked, often marginalised and not necessarily respected for their work. It is not surprising, therefore, that the group shared associated feelings of anger, despair, indignation, frustration, cynicism and sometimes denial. Although the group demurred somewhat at having to acknowledge these concerns and their emotional responses to them, they agreed that this was indeed the context within which any notion of hope needed to be set. How, then, did they conceptualise hope and what sorts of sources did they identify?

It was obvious in the initial interviews that some participants could readily identify with particular sources of hope, while others took more time to name them. Since the autobiographical writing demanded more focused attention, it was often here that ideas began to crystallise. The focus group weekend deepened this process further as the group began to share their experiences and ideas with each other. In the statements that follow I have particularly chosen those which have an environmental focus.

Nathan here reflects on what he learned in childhood, growing up near his grandparents' farm:

> My playground was a beautiful area bounded by rivers, moorland, woods and empty farmland. Both my parents love it, love natural beauty, love wildlife and are great respecters of it. They are as enthralled by the emergence of celandines on a spring morning as some people might be by the birth of a baby. To them it's a source of conversation and delight and this they passed on to myself and my brothers. In our early childhood when we went for walks each weekend, although we didn't know it, we were told what everything was and

> what the behaviour of things were, birds and animals. . . . And we knew all the wild flowers and where they appeared and why, and that I absorbed. It's a magical world in the sense that however close you go into it, it sustains its beauty. . . . And nature never lets you down, it is forever beautiful and, best of all, it provides all our needs. Except love, you have to learn to love.

This description captures both the wonder and awe experienced by Nathan in childhood and the role of his parents in grounding that experience in knowledge about the environment. When work gets difficult, Nathan says, he still retreats to the woods or fields, for nature creates a sense of renewal and puts things in perspective.

By contrast, Henry describes the impact of his first journey outside Europe, travelling in North Africa. Here he is profoundly affected by the desert, by people, by exposure to a non-European culture.

> Six months of the open road with a 20-litre bag and a diet of bread, sardines, tomatoes and oranges. Awareness of the vastness of the open night sky over the desert. Awareness of the life that continues in the desert. Being frightened by guns, alarmed by the lack of disinfectant, being surprised by friendships offered. A temporary period of material hardship and a subsistence lifestyle and an opportunity to share life experiences with people in the Mahgreb. Appreciating community and realising that so many Europeans are themselves marginalised from this experience. . . . Travellers can always tell a story. Stories of hope from across the boundary lines.

Hope can come from awareness of difference, learning from other people's experiences, especially from cultures which give new insights into our own.

Another participant, Emma, writes of the impact of living in the South:

> The year I spent in Vietnam, along with earlier experiences of living in other cultures, continues to provide me with a reference point for measuring what is really important. When you see people coping with privation, with loss or separation, with lives that benefit from few of the luxuries we often take for granted, you are wise to put into proportion your own difficulties and discomforts. When you gain some insights into the things that give those people joy or peace of mind: religious belief; love of the family and the land; music and song; dance; poetry, you have to look beyond the everydayness of your own life to what is at the core.

Beginning to see what is at the 'core' of life deepens one's understanding of the human experience. It puts the everyday in perspective, and hope partly emerges from becoming clearer about what it is that really matters.

Jake, however, who works in many conflict situations, reflects on the darker side of human nature:

> Both as individuals, and in the societies we create, there are deep flaws which it seems can never be overcome. This is a cause for despair, if one is looking for evidence of real change towards a better world. How then to maintain hope? For me, I sometimes think it is genetically programmed. . . . More recognisable sources of hope include: the friendship of colleagues and others, particularly when they recognise and meet my needs before I have articulated them; plans that I have made with others which become reality; laughter and unrestrained merry-making (never enough of this); a conviction (but this is *a priori*, not intellectual) that I am part of a much larger and purposeful whole; meditation leading to a calm spirit and mind.

While Jake is involved with seemingly intractable conflict situations in his professional life, it is from friends and colleagues that he draws his support and therefore his hope. He also draws attention to the role that his spiritual beliefs play in this.

Participants generally identified several sources of hope, sometimes with interesting commentary. Thus Nathan recalls: 'A curious sense of expectation at times of change – a misplaced sense of excitement perhaps, but hope can feed on that. Not that it raises any confidence that our current situation will improve, but it pushes ajar the doors of perception, and I feel more alert to opportunities and trends.' He describes his five main sources of hope, beginning with childbirth, an undoubted 'renewal of hope' in which 'we routinely acknowledge our hopes, dreams and ambitions through our children'. Second, he writes, 'I need the uncaring consistency of nature both as a bedrock to restore my hopes in dealings with people and as an energising source of inspiration.' Then he thinks of 'certain people under duress who never allowed themselves to be overwhelmed by the tragedies that befell them', recalling a man who watched his wife and children die under the Khmer Rouge. Fourth, he remembers 'the generosity and undemanding hospitality of economically poor people in many countries around the world, in their reception of me, a relatively wealthy stranger in their land'. Last comes music, 'from the sound of whales singing to Pachelbel's Canon [which], like art, ennobles human endeavour and complements the human condition at every level'.

Here, as we listen to these environmental and global educators 'thinking out loud', we can begin to distinguish a range of themes relating to hope and inspiration. At the end of the focus group weekend participants were asked to brainstorm the sources of hope that they felt they had collectively identified. These are shown in Table 7.1.

This table is a striking tribute to the resilience and resourcefulness of this group of educators. Although many of these sources of hope were previously implicit in their personal and professional lives, it was the research process itself which made them explicit, visible and therefore more available as a source of creative energy. Thus one of the most striking things about the research process was that the focus group provided not only the richest data but it also became a source of hope in its own right. Participants said they felt nourished, renewed, clearer about what kept

*Table 7.1* Sources of hope (1)

*The natural world* – a source of beauty, wonder and inspiration which ever renews itself and ever refreshes the heart and mind

*Other people's lives* – the way in which both ordinary and extraordinary people manage difficult life situations with dignity

*Collective struggles* – groups in the past and the present who have fought to achieve the equality and justice that is rightfully theirs

*Visionaries* – those who offer visions of an earth transformed and who work to help bring this about in different ways

*Faith and belief* – which may be spiritual or political and which offer a framework of meaning in both good times and bad

*A sense of self* – being aware of one's self-worth and secure in one's own identity, which leads to a sense of connectedness and belonging

*Human creativity* – the constant awe-inspiring upwelling of music, poetry and the arts, an essential element of the human condition

*Mentors and colleagues* – at work and at home, who offer inspiration by their deeds and encouragement with their words

*Relationships* – being loved by partners, friends and family, which nourishes and sustains us in our lives

*Humour* – seeing the funny side of things, being able to laugh in adversity, having fun, celebrating together

them going, more aware of the 'tapestry of hope', witnesses to faith in the human spirit.

## Some conclusions

The *a priori* theory underlying this research was that sources of hope play an important role in the life and work of global educators. As a result of working with this particular group of educators, three main conclusions can be drawn. These are: (1) sources of hope are of crucial importance in the personal/professional lives of these global educators; (2) ten main sources of hope were identified by participants; (3) the research process itself, particularly the focus group weekend, was empowering for participants. Whether such conclusions are applicable to other groups of educators remains to be seen. These are, it should be said, preliminary findings which have focused on *sources* of hope.

Four methods of validation were incorporated into the research: namely, triangulation, face validity, construct validity and catalytic validity (Lather 1991; Hillcoat 1996). First, *triangulation* – that is, the interviews, autobiographical writing and focus group, produced similar and often overlapping material which, taken together, also deepened participants' analyses of hope. Second, *face validity* was achieved in that participants 'recognised' and agreed with the description of the research process and emerging analysis recycled back to them for comment. Third, *construct validity* was present in that the research made visible and explicit what, for many, had been invisible and implicit in their lives. The project's initial theory was thus validated by the experience of participants. A measure of *catalytic validity* was also achieved, with people reporting both increased self-understanding and an energising of their personal and professional lives.

The findings of this research project need wider validation with other groups of educators. Three particular questions which arise are: (1) what are the key variables which orientate people towards different sources of hope? (2) in what ways are such sources of hope socially constructed? (3) how does such a sense of hope relate to praxis and emancipatory action? It is interesting to note that several of the sources of hope identified are themselves currently under threat. Thus the natural world is being seriously damaged on a global scale; for many people the struggles of everyday life actually become more than they can manage; most visions of the future in the late twentieth century are pessimistic or dystopian; and faith in both traditional politics and religion has been generally declining.

The lesson here for educators is that hope has a crucial role to play in how we manage our lives in turbulent times. As Huckle (1990: 159), cited previously, maintains, we must 'build environmental success stories into our curriculum and develop awareness of sources of hope'. No problem, environmental or otherwise, should be taught about at any level of education without concomitant emphasis on positive strategies for its resolution. And this includes, of course, acknowledgement that people differ ideologically over preferred solutions. To educate in a spirit of hope is not to be unrealistic, since, if grounded in socially critical and reflective practice, it becomes a source of empowerment and strength to both teacher and taught.

Deconstructive postmodernism not only leaves us 'between' stories, but without any possibility of a new story. Constructive or revisionary postmodernism, however, rather than seeking to eliminate the possibility of such world-views, looks for a new synthesis. Griffin writes:

> Constructive postmodern thought provides support for the ecology, peace, feminist and other emancipatory movements of our time, while stressing that the inclusive emancipation must be from modernity itself. . . . Through its return to organicism and its acceptance of non-sensory perception, it opens itself to recovery of truths and values from various forms of premodern thought and practice that had been dogmatically rejected by modernity.
> 
> (D. Griffin 1988: xi)

Hope has a central role to play in revisionary postmodernism, not in the shallow sense of hoping that things will improve but in the sense of accessing deep sources of inspiration. Thus McKibben (1995: 5) writes: 'I no longer think fear is sufficient motivation to make [the] changes . . . to spur us on we need hope as well – we need a vision of recovery, of renewal, of resurgence.' In his introduction to *Pedagogy of Hope* Freire writes:

> I do not understand human existence, and the struggle needed to improve it, apart from hope and dream. Hope is an ontological need. Hopelessness is but

hope that has lost its bearings, and become a distortion of that ontological need.
... Hence the need for a kind of education in hope.

(Freire 1994: 8)

Cynicism about hope is one element in the psychology of despair. Huckle's (1990) emphasis on the need for optimism in environmental education is not a turning away from reality but an intuitive understanding of what the human condition most requires. We need, writes Richardson:

> Stories – myths and folktales as well as true accounts – to help us hold the beginnings, middles and ends of our lives together. Without them we shall not have hope: yes, to lose stories is to lose hope, but conversely to construct and cherish stories is to maintain hope.
>
> (Richardson 1996: 101)

Environmental educators are well placed to contribute to such stories, whether on the local scale (Cheney 1989) or the evolutionary (Swimme and Berry 1994). We also need to develop new stories of the future as well as the past – visions of an earth restored. The centrality of the natural world both as a source of concern and a source of hope among participants may go some way to validate Wilson's biophilia hypothesis (Kellert and Wilson 1993), that humans do indeed have an innate affinity for all living things. But then only the tale of modernity could ever have made us forget that.

Chapter 8

# Always coming home
Identifying educators' desirable futures

## Summary

This chapter reports on a research project which facilitated a group of socially committed educators to clarify the nature of their desirable futures and to identify their sources of hope. An account is given of how a participatory and experiential focus group format enabled participants to explore these issues in a fruitful and synergetic way. Key elements of their preferable futures and primary sources of hope are identified and compared with previous findings. Finally, these concerns are located within wider contemporary debates about revisionary postmodernism, sustainable futures and the utopian tradition.

Ursula Le Guin (1988: 1) begins her novel *Always Coming Home* with these words: 'The people in this book might have been going to live a long, long time from now in Northern California.' She describes her task as a sort of 'archaeology of the future', an attempt to discern the cultural contours of a society existing somewhere in the far future. This chapter also reports on such an archaeology, but of the not so distant future. Its contours, by contrast, are drawn from the social imagination of the late twentieth century and in particular a research project on identifying desirable futures.

## Desirable futures

Sohail Inayatullah (1993: 236), in setting out various frameworks for studying the future, notes that 'futures studies largely straddles two dominant modes of knowledge – the *technical* concerned with predicting the future and the *humanist* concerned with developing a good society'. On the one hand, there is the wish to predict the future and, on the other, the desire to create more livable futures. The importance of this normative endeavour has been long recognised within futures studies. Thus Cottle and Klineberg argue:

> Without the ability to dream, to imagine events that have no concrete reality, there can be no creativity in general nor any possibility of creating the images

of future goals that inform present experiences and give directions to current strivings.

(Cottle and Klineberg 1974: 31)

Similarly, Bell and Mau (1971) stress the crucial significance of images of the future in their introduction to *The Sociology of the Future*. So important has this strand of futures become that, twenty-five years later, Bell (1997) devoted a whole volume of his authoritative work, *Foundations of Futures Studies*, to the exploration of preferable futures.

There are many reasons for this ongoing interest in preferable or desirable futures (the terms are used interchangeably here). Thus Dator argues that

> futures studies tries to . . . help people move their images and actions beyond an attempt passively to forecast the future and then to develop plans of action on the basis of the forecasts. . . . The next step is to generate positive visions of the future – to create preferred futures – and to base planning and decisions on them.
>
> (Dator 1996b: 109)

This concern is reflected in the work of many contemporary futurists, such as Hazel Henderson, Barbiera Masini, Martha Rogers, Johan Galtung, Allen Tough and others (Inayatullah 1996). Much of this interest, as earlier chapters have noted, stems from the work of Polak (1972), who argued that popular images of the future act as mirrors in reflecting the health of society. He argues that positive images act as a 'guiding star', drawing societies on to achieve greater things, whereas negative images of the future mark a society in stagnation or decline.

Whether or not the correlation is as simple as this, images of the future clearly play a crucial role in social and cultural change. My own actions in the present, for example, are influenced by the images I have of my past and future 'selves'. If I feel grounded, acknowledged and competent in the present, it is because this image of myself has been reflected back to me by friends and colleagues in the past. I thus project this image of myself ahead into future personal and professional engagements. There is an element of self-fulfilling prophecy here that we have all experienced in relation to teacher expectation during our schooling. Trying to grapple with societal images of the future is a much more complex, but necessary, task.

Chapter 5 identified pioneers in this field, and Chapter 6 showed Boulding's influence on my own research in this area. The study described in this chapter represents a further investigation in this field, an experiential workshop looking at desirable futures and sources of hope together.

Although few futurists have had much to say explicitly about hope, it clearly has a crucial role to play in understanding the human condition. It is hope which releases the energy needed to imagine things being other than they are, to note that which needs changing, to develop some vision of a better situation and to work towards it, even if sometimes at great cost and sacrifice. Hope can be spoken of in both a weak sense – such as hoping that something may come about (whether realistic or not) – or a strong sense, in which the power fuels major visions, such as the ending of apartheid. What is the role of hope, therefore, in identifying and working towards visions of desirable futures?

As described in Chapter 7, the literature on hope is scattered, but it is nevertheless a perennial phenomenon to which commentators have turned their attention. Stotland (1969), in *The Psychology of Hope*, is primarily concerned with the role of hope in relation to anxiety, depression and healing. Desroche (1979), in *The Sociology of Hope*, focuses on hope as a religious phenomenon, particularly in relation to millenarian movements. Bloch (1986), in *The Principle of Hope*, casts his net more widely, drawing on Marxist and Romantic traditions. And Moltmann, in *The Experiment of Hope*, extensively explores hope in spiritual and secular contexts. He writes:

> Whoever begins with hope is aiming to create new experiences. Hope does not guarantee that one will have the wished-for experiences. Life in hope entails risk and leads one into danger and confirmation, disappointment and surprise. We must therefore speak of the experiment of hope. . . . In the experiment of hope the object at stake is always one's own life. . . . Hope must therefore be sufficiently comprehensive and profound. It must encompass happiness and pain, love and mourning, life and death if it is not to lead us into illusion.
>
> (Moltmann 1975: 188)

The question then becomes: what might it look like if we consciously want to explore our notions of hope? The literature on actively working with hope is scattered, although some interesting examples have been gathered together in the proceedings of a seminar entitled *Aspects of Hope* (Carter *et al.* 1993).

## Exploring futures and hope

In Le Guin's story, the Kesh (the people whose lives she is exploring) speak to us through their stories, poems and songs. They act as witnesses from the future, although, as she says, 'All we ever have is here, now.' As a result of my own previous work in this area, I became interested in three questions. What can we learn by bringing together the literature on preferable futures and hope? What happens if we try and actively work on both of these together? What process might best be used in order to do this?

Qualitative research today is in an exciting state of flux. In defining the complexities of this wide-ranging field, Denzin and Lincoln (1994) describe a series of historical 'moments' which qualitative research has gone through, from the traditional and modernist to the postmodern. Researchers can now draw on multiple methodologies in the attempt to make sense of, and interpret, different phenomena in terms of the meanings people give to them. Ethnography today, Denzin argues, is a 'moral, allegorical and therapeutic project':

> The ethnographer's story is written as a prop, a pillar, that . . . will help men and women endure and prevail in the frightening twilight years of the twentieth century. These tales record the agonies, pains, successes, and tragedies of human experience. They record the deeply felt emotions of love, dignity, pride, honour, and respect.
>
> (Denzin 1997: xiv)

Such issues require a research process which is both participatory and experiential. In order to achieve this, the format chosen was a residential focus group which allowed participants to come together for a fixed period of structured time. In the past I had often worked 'cold' with groups when running envisioning workshops; that is, we had not met previously nor had any preparatory work been carried out. However, Jungk and Mullert (1987) point out that people's dreams are often deeply buried because they were trampled on in childhood, so some delicate work may be needed to release them, especially into the public domain. The focus group sessions were thus designed to be experiential, participatory, interactive and cooperative. This allowed a climate of trust to develop, in which participants felt freer to contact and share their deeper feelings. The invitation set out the weekend's concerns as follows:

> As socially committed educators we spend much of our time dealing with an array of problems that have come to be associated with living in 'postmodern times'. There are thus various contemporary issues that we wish to see dealt with in education, ranging from the personal to the political and the local to the global. These may often relate to human behaviours which we wish to challenge and say 'no' to. It is equally important, however, that we spend time and energy on what we wish to promote in society, what we wish to say 'yes' to. In particular, therefore, this research project is concerned with: (a) identifying the nature of our desirable futures; (b) exploring the nature and sources of our hope; and (c) investigating more holistic ways of learning and knowing. The weekend will provide a forum for participants to share, test out and clarify their understandings of both 'hope' and 'desirable futures' with other members of the research group.

The group that eventually came together as a result of the invitational process consisted of seventeen people – ten women and seven men – whose jobs ranged from NGO education advisers and senior lecturers in higher education to project coordinators and freelance educational consultants. Data were gathered through sample interviews, individual journalling and on-the-spot transcribing of small group discussion. Previous research had shown the effectiveness of these processes, although this time more emphasis was placed on journalling during the focus group itself (Progoff 1975).

An extended focus group has several advantages over other methods of data collection (Vaughn et al. 1996). First, it offers direct contact with participants themselves; second, it offers both security and challenge to participants; third, the group format is dynamic and promotes interaction; fourth, discussion between participants generates spontaneity and enthusiasm; and fifth, a synergy can occur which leads to richer and more multilayered data. All of these fruitfully occurred during a weekend at a quiet residential centre on the edge of the Cotswold hills. It was a location where I knew busy people could relax and unwind in order to be present to the issues at hand.

The dilemma of the ethnographic researcher is constantly one of 'voice' – how to report on so much that is written and said; how to capture the flavours of the event; how, in this record, to capture the moments of insight that participants experienced. What is offered here is a gestalt of the weekend returned to participants for their comment, amended accordingly, and thus 'recognised' by them.

The process of reflection on desirable futures began before the focus group when participants were asked to record their preliminary thoughts on the event. Not everyone found this easy to do. As one person commented: 'I have a real difficulty about looking at the future because I don't want to be depressed and I can only see painful futures.' However, this 'warming-up' generally bore fruit in the first phase of Boulding's (1994) process, which requires individuals to draw up a wish list of the changes they would most like to see prior to the commencement of their 2025 envisioning. After making a record of their visualisation, participants worked in pairs to produce self-explanatory posters depicting the key features of their preferable futures. This meant that three sources of data – wish lists, envisioning notes and posters – were available to draw on in establishing the nature of the group's desirable future.

In looking at the main changes people wanted to see, eight key areas of concern were found in the wish lists: the nature of community; the hazards of consumerism; the quality of the environment; elimination of poverty; greater equality in relation to gender and human rights; less prejudice between individuals; a reduction in violent conflict; and greener transport. These issues marked out the broad terrain for the weekend.

As always, the envisioning process (see Chapter 6) tapped deeper layers of knowing and often transformed intellectual ideas into powerful visual images. Some participants found themselves in rural areas, others in towns or cities. It was as if each participant's subconscious showed them particular glimpses which added flesh to the bones of their desirable future. However, because these images are like snapshots rather than movies, they give a selective picture of the terrain. Four main themes emerged from the envisioning: the nature of community, the quality of life, altered townscapes and the quality of the environment. These and the descriptors used by participants are listed in Table 8.1. There is obviously an overlap between these four broad categories.

After working in pairs to produce posters depicting the key features of their desirable futures, the group mounted these on the wall and then examined them for

*Table 8.1* Social educators' preferred futures 2025

*Conviviality* – calmer pace of life; less stress; smiling, energetic people; time to talk; more joy but also sadness; richer quality of relationships; no rush, people relaxing; comfortable and colourful clothes; lots of laughter

*Community* – locally produced goods; more jobs based at home; doors open, no burglar alarms; recycling schemes; groups open and welcoming to others; bartering and skills exchange; ease between the generations; digitally interactive notice boards for voting on local issues; unhurried and more reflective people; music and street theatre; deschooling

*Towns* – human scale; clean and healthy; trees, gardens, fountains; energy-efficient buildings; renewable energy; no supermarkets; small shops and market stalls; vibrant cultural centre; multicultural; people and children in spacious streets; sculptures, frescoes and public spaces; absence of mechanical noise; bustle of activity; no beggars or homeless; no cars; bike routes, trams and trains; organic gardening; easy access to countryside

*Environment* – a green manifesto; sunshine; birdsong, clean air, calm and beautiful countryside; forests, valleys, hills and lakes; flowers and animals; brightness and light; closer relationship with and greater respect for nature

common elements. In particular, they were asked: what is the essence of these visions? Participants described this in various ways, each capturing a crucial facet of the collective vision that was emerging. Among the most striking were the following:

- 'Valuing of self, others and the earth.'
- 'The integrity of creation.'
- 'A paradigm shift in cultural values.'
- 'A new set of assumptions about human coexistence, emphasising sustainability and interdependence.'
- 'The social pathologies of multiple oppression have been healed.'

Prior to the focus group weekend, participants had also been asked to record their initial thoughts and feelings about the importance of hope in these times. Their reflections included the following:

- 'Sometimes I think that hope is all we have left.'
- 'Hope, I think, is what makes order out of disorder.'
- 'There is something about hope that feels passive.'
- 'Hope liberates, motivates, enables, strengthens, gives joy a foothold.'
- 'I don't want to look at hope because it is too insubstantial.'
- 'Hope and joy as a life position, in the full knowledge of life's agonies.'
- 'For me the importance of hope in difficult times is central.'

During the weekend, some of these initial observations were to be reinforced and others discarded.

In order to facilitate the subsequent discussion on this theme, participants were asked to bring with them an object, picture or piece of writing that was a symbol of hope for them. Then, in small groups, they took it in turns to talk about the objects that they had brought. Since these were often of great personal significance, the stories shared were listened to attentively and respectfully. As people took it in turns to speak, the objects were laid ceremonially in the centre of the circle. Participants found this an absorbing experience that affected them in different ways.

- 'Sharing our sources of hope was like . . . walking through a door into another person's heart.'
- 'Hope is the feeling of energy that flows through me when I listen to people sharing their inspirations.'
- 'Re-minding ourselves of the wholeness that precedes fragmentation, and the wholeness that begins again through the telling and sharing of our stories.'
- 'Our group are hunters of culture and communal experience; it's a wild, indomitable group with more hope than fear, confident, articulate, warming, connecting.'

Posters were also used for each group to summarise what members felt they had learned about their sources of hope. As a result of this process it was possible to identify nine main sources. These are summarised, with their descriptors, in Table 8.2.

At the end of this exercise the group was faced with the question, what is the essence of your hope? This was described in different ways by participants, including the following:

- 'Hope is what moves you from the dark place.'
- 'All that sustains engenders hope. All that destroys and harms engenders hopelessness. There was an interesting discussion in our group about hope having to embrace the light and the dark.'

*Table 8.2* Sources of hope (2)

*Roots* – links with the past; childhood; history; previous generations; ancestors; the need to honour continuity

*Human creativity* – individual and communal; music, song and dance; painting and sculpture; utopias; books, stories, poetry

*The natural world* – constancy in nature; rebirth in the land; continuity and timelessness; the season; animals; regeneration

*Collective struggles* – for justice; solidarity; people who suffered or died for a cause; networking and activism; people working together

*Other people's lives* – amazing people; those who do their own thing; children – now and in the future; heroes/heroines

*Relationships* – families, friends and loved ones; gifts given by them; all those who support us

*A sense of self* – self-knowledge and expression; the potential of myself; stepping outside of myself

*Faith and belief* – religious and spiritual traditions; the politics of daily life; new ideas

*Humour* – the importance of laughter and satire; being able to laugh at oneself

- 'The essence of hope must, then, be the faith in each other and an ability to be with each other. To recognise our capacity for good (and for evil) and to help each other to create the climate for us to let down our defences and fears. Fear is the essence of no-hope. Open acceptance is the essence of hope.'

Finally, as the weekend progressed, some participants began to make connections between what they had learned about their desirable futures and their sources of hope.

- 'At the heart of our vision is hope and a sense of shared values. . . . The key to hope and hence the practicality of our visions, the essence, is about the way we are with each other – here.'
- 'Sharing our sources of hope was beginning to connect with the particular individual life experiences that for each person help to piece together a vision of a different future; that are the grounding and material of that vision.'

## Creating communities of hope

Although it is not appropriate to make generalisations from the data described here, it is possible to compare these findings with previous research. This can be done in relation to (1) the nature of desirable futures; (2) sources of hope; and (3) the impact of the research process. This research, though small-scale, extends what we already know about the efficacy of envisioning. The results described above show again, as does Boulding's (1994) more extensive work, that people can envision hopeful futures given an appropriately facilitative environment. The main types of imagery she found in her workshops have already been described in Chapter 6. I believe that if we look at the features listed in Table 8.1 above – that is, the main characteristics identified by this group as significant in their preferable futures – they would not look out of place in Boulding's 'baseline' future. Similarly, the findings reported in Chapter 6 on students in higher education also showed that their preferable future was green, convivial and peaceful, with an emphasis on alternative forms of transportation. Thus, despite a significant age difference, there is considerable similarity between the preferable futures of both these UK educators and students, which echo Boulding's wider findings from the United States and elsewhere.

This research also adds to what we know about people's prime sources of hope. Of the nine main categories shown in Table 8.2, eight are the same as those identified in the study described in Chapter 7. The two categories missing on this occasion are 'visionaries' and 'mentors and colleagues'. On the other hand, an additional category of 'roots' has been identified. This focus group weekend was also designed to promote wider discussion on participants' notions of hope and, as the comments above indicate, this began a rich process of reflection. It is probably impossible to disentangle personal and professional sources of hope, although it would be interesting to see if the same ones came up under a specifically personal rubric. The point to note, of course, is that both this group of educators and those described previously had in common a professional interest in teaching and learning about social, political and environmental issues. These sources of hope may thus only be characteristic of such educators, although I suspect that all futurists with a normative orientation will be able to relate to them in some way.

How did people respond to the experiential and participatory process used in this focus group? It was designed, as previously stated, to develop a deeper climate of trust, sharing and reflection than might otherwise have occurred and, indeed, participants commented on this in their journals. Thus, before coming they were asked what their hopes and fears for the research weekend might be. Among their hopes were the opportunity to: slow down, find new allies, be affirmed, reformulate goals, and renew energy and commitment. For a group of experienced educators, their fears were interesting: feeling isolated, being out of one's depth, feeling shy, not contributing anything useful, the potential of the group not being realised. Journal comments accordingly reflected both the insights that people gained through the group sessions and their responses to each other and the process more broadly.

At the end of the weekend participants were asked if the research process had in any way affected their learning. Responses included the following:

- 'The interaction in our group was a source of hope.'
- 'The recognition of a shared human-scale conception of the future gives energy for its realisation.'
- 'Amazing how similar our visions were.'
- 'Recognition of the need to work through difficulty.'
- 'The weekend itself was a source of hope.'
- 'The value of places like this where communities of hope can come together.'

As these comments indicate, the research process was seen as an invaluable experience in itself. Most people's prior fears were allayed and many of their hopes for the weekend were met. More importantly, this shows how a participatory process can enhance the quality of the data gathered and, in this case, how it can be a major source of hope in its own right. As all person-centred educators know, the medium is the message.

## The wider picture

It is also possible, and important, to locate research such as this in the wider socio-cultural context. This is done here in relation to debates about (1) the nature and significance of postmodernity; (2) the notion of a more sustainable society; and (3) the utopian tradition.

Various commentators, in trying to make sense of the complexities of post-modernist thinking, have made an important distinction between deconstructive or sceptical postmodernism, on the one hand, and constructive or revisionary post-modernism, on the other (Rosenau 1992). Thus, whereas the former is distrustful of all metanarrative and wishes to deconstruct totally the 'story' of modernity, the latter, while accepting the critique of scientific rationalism, wishes to construct a re-visioned story that goes beyond the fixity of modernity. Griffin writes:

> Going beyond the modern world will involve transcending its individualism, anthropocentrism, patriarchy, mechanisation, economism, consumerism, nationalism and militarism. Constructive postmodern thought provides support for the ecology, peace, feminist and other emancipatory movements of our time, while stressing that the inclusive emancipation must be from modernity itself. The term postmodern, however, by contrast with premodern, emphasises that the modern world has produced unparalleled advances that must not be lost in a general revulsion against its negative features.
> (D. Griffin 1988: xi)

The study of desirable futures can be seen as an integral element in this process of re-visioning. Indeed, Walker (1996: 67), in his exploration of postmodernism and the future, argues that 'using a postmodern paradigm for the futuring process creates significant advantages for futurists'. These include its flexibility and its stress on the use of imagination and intuition to inform perception and reason. Normative futures are thus able both to benefit from the insights of postmodern thinking and to contribute to the emergence of a rich revisionary postmodernism.

The findings from this and similar research about people's desirable futures can also be firmly located within the debate about the nature of a more sustainable society. Indeed, the futures described here all provide glimpses of what a more just and ecologically sustainable future might look like. Whether the envisioners consciously realise this or not, these are classic pictures – for example, of Milbrath's (1989) 'new environmental paradigm', Dobson's (1995) 'sustainable society' and Trainer's (1995) 'conserver society'. They typify much, but not all, of the visionary writing about sustainability which tries to envisage the changes in lifestyle that would be needed to bring such a society about. It abides by green principles, emphasises the local, is on a human scale, is more personally fulfilling, rich in ends and simple in means.

It appears, therefore, that when interested people are given the opportunity to envision their preferable futures they often come up with forms of social and political organisation which are less materialistic and more ecocentric in perspective, as witnessed by work with Local Agenda 21 programmes, such as Vision 21 (1997). This would support Milbrath's (1989) contention that, in the late twentieth century, we were seeing a clash between the dominant social paradigm based on modernity's mechanistic world-view and an emerging paradigm which is based on more holistic and ecological principles. Some aspects of such a shift have been mapped by Worcester (1994) in the United Kingdom and, more widely, by researchers such as Inglehart (1997), with his thesis of an emerging post-materialist society.

There is within Western society a long tradition of utopian thinking which has manifested itself in literature, political thought and intentional lived communities.

Such thinking commonly involves a criticism of current society and its ills, together with wide-ranging proposals for their resolution in the form of a radically new society (Kumar 1991). Writers such as Thomas More, William Morris and Charlotte Gilman Perkins thus described societies which were very different from those that they lived in. Such accounts drastically challenged the mores of the time and offered readers a vision of how things could be radically different. Traditionally, such utopias have been located in unknown lands, far distant pasts or, more recently, distant futures and other galaxies.

Many ecological thinkers have noted how much at home they would feel in William Morris's *News from Nowhere*, even though it was written just over a century ago (Coleman and O'Sullivan 1990). Since then, for a variety of historical reasons, the twentieth century has largely been a place of lived and literary dystopia rather than utopia. This began to change in the 1970s, however, as the tradition was reworked by those with an interest in ecological and feminist utopias. Piercy's (1976) *Woman on the Edge of Time*, Callenbach's (1981) *Ecotopia Emerging* and Starhawk's (1993) *The Fifth Sacred Thing*, each offers accounts of what needs to be opposed and visions of what might be created instead. Sargisson's (1996) *Contemporary Feminist Utopianism* offers valuable comment on many of these issues and debates.

The educators described in this research, once given the opportunity to carry out their own excavations of the future, clearly articulate the hopes and fears of these times. They reveal the contours of a future which is more convivial, just and ecologically sustainable. They recognise too the experiment of hope – 'hope and joy as a life position, in the full knowledge of life's agonies'. Having come together and seen into each other's hearts, they went away refreshed and strengthened in the knowledge that they were not alone.

Desirable futures, sources of hope, revisionary postmodernism, a more sustainable society, the utopian tradition – all have much to contribute to our thinking in these times. Each can significantly sustain us in our personal and professional lives. Beginning to explicate the significance of each and to interweave them is one of the deep socio-cultural tasks required at this time in history. Kumar, an incisive commentator on the end of the century, observed that

> What we seem to have today is the apocalyptic imagination without hope but also, more strikingly, a kind of millennial belief almost entirely emptied of the conflict and dynamism that generally belongs to it. It is a millennial belief without a sense of the future. We have, it seems, at the end of the second millennium achieved the millennium, the hoped-for state of peace and plenty. But it brings no pleasure, and promises no happiness.
>
> (Kumar 1995b: 205)

The new century, says Kumar, should provide the element of hope and utopia the element of desire. We are thus 'always coming home' because of the longing for a time – whether in the far past, in the womb, in this life, in a future society or after death – in which our deepest human yearnings will be met. This is at the heart of human striving and if we are fully alive we have no choice but to dream. We should do this most seriously and joyfully both for our own self-actualisation and for generations yet to come.

Chapter 9

# Living lightly on the earth
## A residential fieldwork experience

### Summary

Students studying global issues at Bath Spa University College have often undertaken a residential field week at the Centre for Alternative Technology (CAT) near Machynlleth in mid-Wales. This chapter sets out the background to this experience and gives a brief account of the educational services offered at CAT. This is followed by an account of how the field week was structured, with its three-fold focus on: (1) learning about appropriate technologies; (2) developing a sense of place, and (3) experiencing community living. Finally, student comments on the field week illustrate their evolving responses to this experience.

Having ascended the water-powered cliff railway that takes us from the car park up to the main site, we stand on the cantilevered deck of the upper station looking out across the hills and autumn colours of the wooded Dulas valley below us. The branches on the silver birches are whipped by a cold wind blowing in from the Dovey estuary as we turn to walk across the old slate workings that are now the site of one of the best known eco-centres in Europe. Later, having been given a guided tour of the site, we retire to our self-catering cabins where we split and stack the wood that will later keep us warm. The annual field week at the Centre for Alternative Technology has come round again.

During the 1990s, global futures was a multidisciplinary subject located within the Faculty of Applied Sciences at Bath Spa University College as part of the institution's modular scheme. The subject had two main axes: spatial and temporal. The spatial axis was concerned with the nature and effects of globalisation and understanding the interrelationships between local and global issues, especially those to do with environment and development. The temporal axis was concerned with the historical origins of environment/development issues, various aspects of the current debate about the nature of sustainability, and ways of both avoiding unsustainable futures and of working towards more just and sustainable futures in the local/global community.

The field week described in this chapter provides an example of how students gained practical experience of simpler ways of sustainable living. Prior to the field week, students had explored the nature of futures studies, debates about the nature of postmodernity (Walker 1996; Littledyke 1996) and the need to be able to envisage preferable futures, including those that embody indicators of sustainability (Trainer 1995; Douthwaite 1996). These themes offered a theoretical underpinning for their visit to the Centre for Alternative Technology in Powys, mid-Wales.

## The Centre for Alternative Technology

In 1974 a derelict slate quarry in the Dulas valley, just north of Machynlleth in mid-Wales, became home to a small group of idealists who were inspired by the challenge of building a living community in which the emerging alternative technologies of the time could be put to the test in a practical way. Originally, the Centre was conceived of as an experimental, largely self-sufficient community, but interest in both this and the innovative technologies being used led to the Centre being opened to the public. Its 40-acre site now has working displays of wind, water and solar power, low-energy buildings, organic growing and alternative sewage systems. In 1990, the Centre became a public limited company in recognition of the need to respond to a growing public interest in environmental issues, and CAT offers practical ideas and information on environmentally sound practices to organisations, students and the public as the major visitor centre in mid-Wales (CAT 1995).

In addition to the resources on the site itself, the Centre offers an information service on all aspects of sustainable technologies. It produces more than eighty publications, which together provide a comprehensive guide to sustainable living. Information about courses and publications are available via the CAT website (www.cat.org.uk). The Centre offers a professional consultancy service to advise larger projects requiring specific expertise – for example, in relation to project design and assessment. Much of CAT's work is also particularly relevant to Local Agenda 21 initiatives.

The education department receives 25,000 school-age visitors each year and produces materials for pupils and teachers to use on site and in schools. Residential visits and professional development services are also available. The Centre offers an extensive programme of more than 120 courses each year, including a highly respected range of workshops on renewable energy, self-build, water treatment and organic food growing.

Just off the main visitor site are two eco-cabins which provide the main residential accommodation. They are built in the timber-framed style pioneered by Walter Segal and are well insulated, with turf roofs and double glazing. They are designed for educational groups to 'experience' sustainable living at first hand. In close contact with each other, students learn how to monitor and manage their own electricity generation, wood and water consumption, and sewage recycling.

The cabins are isolated from mains electricity, so that all the power used must be generated at the cabins. There are separate 24-volt batteries for each cabin, which are charged by four energy sources: wind generators, a mini hydro-electric scheme, photovoltaic panels, and a diesel generator for emergency use if renewables run out. The use of these energy sources can be monitored by an instrument panel in each cabin, together with the total energy input and consumption.

The cabins are heated by wood-burning stoves, the wood coming from local forests which are regarded as a renewable source. The stoves may be used for cooking, but gas cookers are also provided. Wood consumption is monitored by weighing the wood before burning, while gas consumption is also metered. The water supply for each cabin is from its own large tank which can be filled automatically, or, alternatively, students can fill the tanks with buckets from a nearby tap. By keeping a record of the buckets used, it is possible to calculate how much water is actually used. The cabins have flush toilets, but a compost toilet is also provided, the use of which considerably reduces the number of buckets of water which need to be carried. Each flush of the toilet uses about two buckets.

## The field-week experience

The facilities provided at the Centre offer an intensive experience of study and communal living. The field week provided an opportunity for students to get to know each other, and it became a shared experience to which they referred throughout the rest of the course. The field week had three main purposes. These were: (1) to carry out practical scientific investigations into appropriate technologies, using equipment and expertise not available at the university; (2) to explore the vision of the future that inspires CAT and the sort of place that has been created as a result; (3) to experience one version of 'living lightly on the earth' in a communal and cooperative context.

On arrival at the Centre, students were given a tour of the site and an introduction to the eco-cabins before joining in a community wood-gathering activity which began the process of working together. This continued throughout the week through the shared responsibilities of buying food, preparing meals and washing up, maintaining the water and wood supplies to the cabin and monitoring electricity use to avoid a power cut. The outline programme is shown below.

### *Monday*

- Guided tour of CAT site
- Introduction to the eco-cabins
- Community activity: fuel wood supply
- Lecture: Introduction to the Centre

### *Tuesday*

- Workshop 1: Efficiencies of hydro/wind/solar generation
- Workshop 2: Organic horticulture and soil analysis
- Lecture: The power of place

## Wednesday

- Group activity: Spirit of the place
- Excursion to Aberdyfi
- Group presentations on place

## Thursday

- Workshop 3: Making and testing a solar panel
- Workshop 4: Analysis of effluent from reedbed sewage system

## Friday

- Review of eco-cabin date
- Depart Machynlleth

The programme of workshops, lectures, student presentations and communal living focused attention on various aspects of sustainability which, taken together, often had a major impact on participants.

The first main purpose of the week was to investigate the operation of selected appropriate technologies. Students took part in four workshops run by Centre staff. The size of CAT's facilities meant that the scale of these investigations was much more realistic than would be possible at school or university. Living in the eco-cabins and monitoring their energy use also provided data for investigation. A brief description of the workshops follows.

## Energy conversions

Measurements were made to calculate the efficiency of both the mini hydro-electric power scheme for the cabins and that of the large system for the Centre (Boyle 1996). The cabins have their own small reservoir which can be discharged through a turbine when required. Measurement of the water flow rate and the height of the reservoir above the turbine allowed the potential power available to be calculated, while a simultaneous reading of the instrument panels in the cabins allowed the actual power being generated to be recorded. Similar calculations were done for the large hydro system on site. A comparison of the efficiencies enabled discussion on why there were differences to take place, raising issues of loss of energy due to friction in the pipes, turbine design and so on.

The efficiency of the photovoltaic panels used to charge the cabin batteries was also estimated. This involved simultaneous measurement of solar insolation, current and voltage being produced and the area of the panels. The efficiency of the wind generators supplying the cabins was also calculated. This requires an anemometer reading to find wind velocity, an estimation of the area swept out by the wind turbine blades, and the amount of electrical power generated.

### Reed-bed sewage system

The site is not connected to the usual form of sewage disposal but has a large reed-bed system for dealing with all the waste from the main site and a smaller system for the cabins (Weedon and Light 1995). The technology of this increasingly attractive procedure was readily taught from inspection of the systems, and measurements taken of turbidity and dissolved oxygen levels at various points in the cabin system. These showed the great improvement in water quality as it passes through the system, and by the end of the treatment it can be safely discharged into the river.

### Solar water heaters

In an attempt to show students that individuals can make steps towards living in a more environmentally friendly way with little expense, they were given the opportunity to construct a simple solar water heater (Trimby 1995). It was encouraging to see the way in which students who had no prior experience of techniques such as soldering or sawing soon gained confidence when undertaking this task. The solar panels constructed were taken back to the university, where their efficiency was measured in a later module on energy.

### Organic horticulture

Although the Centre is situated on a barren slate quarry, it now produces considerable quantities of organic fruit and vegetables, which are used on site in the restaurant. A fertile soil has been created by careful use of composted organic waste, including residues from the compost toilets and the reed-bed systems. Displays illustrate the various composting methods available, both for large- and small-scale applications. Students were shown a variety of techniques for increasing the rate of composting and many came away determined to try some of these methods at home (Dudley 1991).

The second main purpose on the field week was to facilitate an engagement with place – that is, the actual site of the Centre in a disused quarry and the wider mountainous environment. Both are strikingly beautiful, and it was the sheer impact of the place, especially in good weather, which always played a major part in the success of this week. However, developing an appreciation and deeper sense of place was not left to chance. Thus, on the second evening a lecture explored ways in which contemporary Western society has often cut itself off from any direct appreciation of the natural world.

The differences between anthropocentric and ecocentric perspectives on the environment were highlighted (Sessions 1995), and also the argument put forward by eco-psychologists (Roszak *et al.* 1995) that the alienation found in industrialised societies today may well have at its root the repressed pain of our separation from nature. Wilson's biophilia hypothesis (Kellert and Wilson 1993) was also referred to and, in particular, research evidence which shows that people are less stressed and heal more quickly when in natural environments. Sleeping on these ideas helped set the tone for the main activity the next day.

The task here was for small groups to explore the site in order to make some record of the 'spirit of the place' as they experienced it. They had four questions to consider:

1   What does this place *say* to you?
2   What catches your *imagination*?
3   What lies at the *heart* of this place?
4   How can you capture its *essence*?

For the first hour this was an individual activity, and students were instructed to find some place which attracted them where they wished to be alone. They were asked to slow down and see what the place said to them if they 'listened' and to record this in some way. Then this became a cooperative venture in which each group had to weave together and synthesise their individual insights. In the evening, the groups made a creative presentation on what they perceived the 'spirit of the place' to be. Their comments included:

> Here people are living an experimental lifestyle within their everyday lives; they are idealistic but realistic. The mountains and rivers allow the imagination to run wild.

> This place says we have the opportunity for new adventure, it awakens the pioneering spirit. There is a sense of cleanliness and rejuvenation. The place speaks volumes about hope, it rewards us with a sense of our own empowerment, our imaginations fired by practical applications. I feel more forgiving of myself as if I've found my niche; to be here makes the difference.

> People come up on the cliff railway and see a glimpse of the future. There is knowledge and expertise on tap. People here stand up for what they believe in. There is a coexistence here between nature and humans.

Each year the presentations showed that as a result of this exploration students made a much deeper connection with the place, and hopefully other places, than they might otherwise have done.

Living in the cabins contributed to the third main purpose of the field week. Each cabin can sleep up to eighteen people and is self-catering, with a communal space adjacent to the kitchen area. A wood-burning stove both warms the cabin and supplies the hot water. Before arriving, the group was split into twos or threes, each small group being responsible for providing breakfast, an evening meal and lunch over a 24-hour period. Living and eating together amicably in such close proximity required the development of good communication and cooperative skills. The group gradually became aware of the importance of such skills in the effective functioning of the cabin, as they found that not everyone always pulled their weight equally.

Over the short period that they were together students therefore gained a taste of both the benefits and difficulties associated with the creation of community. Apart from the specific tasks that had to be done, this element of the field week was not specifically prepared for in advance. Rather, the students began to learn from their immediate experience about the benefits and dilemmas of communal living. Being able to work cooperatively together and to create a sense of community is, it can be argued, a vital element in any notion of a more sustainable society (Metcalf 1996). A subsequent lecture in the module looked at issues relating to the setting up of intentional communities.

## Student responses

As part of their work for the week, students were asked to keep a journal in which they recorded their responses to the workshops, lectures and other experiences each day. Some flavour of how students responded to the three main strands of the field week is given by the comments below.

In commenting on the *workshops*, many students were struck by the simplicity of much of the technology at the Centre and how applicable it was to other, much wider contexts. However, the biggest impact was in relation to the understanding they had gained of how particular technologies worked in practice and their minimal environmental impact. One wrote, 'It's amazing that we ever used fossil fuels in the first place when there is such an abundance of clean fuels to be used.' And another: 'It was fascinating to see that there was a more natural way of disposing of effluent than constantly pumping it with chemicals.'

All students expressed a very positive appreciation of the Centre and the *place* itself. CAT was described as visitor friendly, with excellent information available for visitors and staff who explained things clearly and who were themselves often an inspiration. One commented, 'All was explained so simply, it was inspirational, I want to start living some of these ideas.' And another: 'I really admire the people there, they showed a possible, cleaner future. That you should only take from nature what it's prepared to give you.'

Many comments on the quality of the location were expressed in single words such as 'wonderful', 'idyllic' or 'beautiful'. Many were affected by their daily proximity to the natural environment: 'The river affected me a lot, I felt in harmony'; 'With nature all around it seemed even more important to value it'; 'The tranquillity merges well with the theme of alternative technology.'

What struck students most about the *community* experience was how they had come to know the rest of the group much better. This included having to adjust to the habits of others and interacting with students and staff with whom they might otherwise not have spent time. Some were struck by the varying degrees of responsibility taken by students in contributing to communal tasks: 'I have very mixed emotions about this aspect. It was apparent that some have the knack of sitting doing nothing. They seemed to lose the sense of why they were there.' All commented that they had been much better fed and warmer than if they had been back in their digs in Bath!

The Centre for Alternative Technology provides an inspiring vision of the future. It has held to the ideals of the original pioneers but has also kept pace with the needs of the times. CAT's rural location adds to its attractiveness, but the principles and technologies it espouses are applicable anywhere. As one group aptly commented in their presentation, 'Appropriate technologies have one foot in the past and one in the future.'

Meadows *et al.* (1992), in arguing the need for visioning, stress the need also to dwell upon our most uplifting dreams. We need too, says Elgin (1991: 77), 'a simple and compelling story of the future', and this is why places like the Centre for Alternative Technology are so important for the twenty-first century. They have the power to encourage transformational processes in individuals, groups and cultural systems. A field week such as this can make a deep impression on students interested in global issues. 'I could never have imagined a place like this running itself so efficiently,' wrote one. 'The only limit is your imagination. It is about what might be, and what could be, if we wanted it to be.'

Chapter 10

# Teaching about global issues
## The need for holistic learning

## Summary

Over the last twenty-five years, many educators have stressed the need for students to learn about global issues and have often taken this to be a relatively unproblematic area of pedagogy. An important Canadian study on the impact of teaching about such issues, however, suggests that the learning process may be much more complex than was previously assumed. This chapter reports on a research project that monitored student responses to learning about global futures and found much more going on beneath the surface than meets the eye. It would seem that strong cognitive, affective and existential responses need to be recognised and acknowledged as part of any journey towards personal and political change.

Teaching about 'global issues' is shorthand for a variety of concerns that an increasing number of educators feel it important to deal with in schools and establishments of higher education. The issues studied range from those to do with the environment, development and human rights, to peace and conflict, race, gender, health and education. They may be part of existing courses in geography, social studies or humanities or take place under headings such as environmental education, global education and futures studies. All of these issues highlight *major*, maybe unresolvable, problems about the human condition. They all involve much human pain and suffering (as well as joy and well-being) and also often impact disastrously on non-human species. Learning about global issues is thus potentially a traumatic activity, so what do students actually experience when they learn about them? And what responsibility do educators have to find out about this?

## Facing global issues

Life at the beginning of the new century – whether personal or professional – is complex, chaotic, fragmented, exciting, challenging and stressful. The hazards of life under late modernity or postmodernity have been well documented and provide the wider background to this study. Giddens writes:

> The crisis-prone nature of late modernity . . . has unsettling consequences in two respects: it fuels a general climate of uncertainty which an individual finds disturbing no matter how far he seeks to put it to the back of his mind; and it inevitably exposes everyone to a diversity of crisis situations . . . which may sometimes threaten the very core of self-identity.
>
> (Giddens 1991: 184)

This feeling of living on the edge, augmented by the local–global paradoxes and ambiguities of globalisation, can lead to an underlying sense of personal insecurity which in turn triggers deeper existential anxieties. Beck (1998) highlights the dilemma of having to make decisions about risks about which we know nothing, and points out that society has become a laboratory where no one is prepared to take responsibility for the outcome of social, scientific and technological experiments.

It is not surprising, in the face of such disorientating change, that many people prefer not to know. 'Most people', says Walsh (1992: 63), 'experience great difficulty in acknowledging the true state of the world, its suffering and its peril. Repression and denial play major roles in this difficulty.' As in our personal lives, some problems are too much to bear so that we deny their very existence. Our defence mechanisms can thus lead to a 'psychic numbing' which denies the pain of the world and our part in producing it. By denying its existence we perpetuate it. Learning about global issues can never be solely a cognitive matter, although many educators may wish it to be. This is why the work of activist academics like Macy is of such importance. In outlining the theoretical underpinnings of her work, she writes:

> Our experience of pain for the world springs from our inter-connectedness with all beings, from which also arise our powers to act on their behalf. . . . Unlocking occurs when our pain for the world is not only intellectually validated, but experienced. . . . When we reconnect with life, by willingly enduring our pain for it, the mind retrieves its natural clarity.
>
> (Macy and Brown 1998: 59)

This echoes the argument of eco-psychologists like Roszak *et al.* (1995) that much of the Angst of the late twentieth century is due to our being cut off and alienated from the natural world in which humans have been rooted for countless millennia.

For those interested in teaching about global issues, the crucial question is to what extent educators have addressed these matters. Certainly, some of those involved in environmental education and futures research have identified the distress and alienation that many young people feel about the human condition. As adults, educators have had longer to live with this awareness and may be better able to inure themselves to it. Young people come to it with a freshness that makes it even more painful. Whereas UK pupils are reasonably optimistic about their own futures, they are pessimistic about global futures (Chapter 3), and Australian students seem to be even more pessimistic (Hutchinson 1996; Eckersley 1999). In their exploration of young people's environmental attitudes in Australia, Connell *et al.* (1999) found that environmental problems made them feel frustrated, sad and pessimistic. Although scepticism and despair among the young are sometimes belittled by adults, they might better be seen as a valuable early warning system for present and future generations.

One of the most comprehensive examinations of the impact of learning about global issues are those carried out by Rogers (1994; 1998; Rogers and Tough 1996). It was this study which provided the inspiration for the research project described here. In her doctoral thesis, Rogers examined in detail the impact on adult education students of a course on global futures taught at the Ontario Institute for Studies in Education. She was able to examine both the personal effect this had on students and the complexities of the learning process itself. She notes:

> Coming to grips with the complexity of the world's problems, confronting uncertainty about the future, and critically examining deeply held worldviews may cause emotional and existential turmoil. To try to cope with the onslaught of thoughts and feelings, people may resort to using defence mechanisms such as denial, suppression, intellectualisation or projection. Consequently, rather than being truly able to face the future, the protective defence mechanisms may cause people to retreat or disconnect from reality. Thus paradoxically, the learning process may lead to paralysis rather than mobilising informed choice and action.
>
> (Rogers and Tough 1996: 492–3).

As a result of interviewing course participants, Rogers observed that five, often overlapping, dimensions or stages of learning were involved (1996) as shown in the box below.

From this detailed analysis of learning about global futures it can be seen that the endeavour is a complex, holistic and deeply personal one – it is far from being solely a cognitive endeavour. As a result of this investigation, Rogers developed the conceptual model of learning shown in Figure 10.1. This highlights the fact that

---

GLOBAL ISSUES: DIMENSIONS OF LEARNING

*Cognitive dimension*

The first stage involves learning new facts, ideas and concepts about the current global situation and its likely future consequences. This is traditionally considered to be the core of teaching about global issues. Some students, when faced with thinking about the future, thought it was 'out-of-touch' or 'airy fairy'. Being forced to step outside their usual spatial and temporal orientations sometimes lead to resistance. Students also felt cognitively overwhelmed, confused and pessimistic when faced with the complexities of the world's problems.

*Affective dimension*

Learning about global issues also involves an emotional response. This appears to occur when knowing shifts from being something intellectual and detached to a personal and connected knowing. Some students experienced a range of conflicting emotions, such as: elation/depression, hopefulness/hopelessness, fear/courage, sadness/happiness. Rogers and Tough (1996: 493) note: 'Grieving has been reported as a common response to learning about global threats to human survival and with respect to human processes of transition and change.' Most importantly, the emotional responses experienced by students need to be accepted and seen as part of a shared experience.

*Existential dimension*

Although rarely discussed in the literature, says Rogers, learning about global issues and possible futures can also lead to a deep soul-searching. For some students this involved a questioning of their values, life purposes, faith and ways of living. They wanted to 'find an answer', or to 'do something', but often found themselves searching without finding any immediate answers. At this level they were being faced with a reconstruction of their own sense of self, something which often occurs when embarking on a quest for deeper meaning and purpose in life.

*Empowerment dimension*

If this upheaval of the soul can be satisfactorily resolved, students can begin to feel a sense of personal empowerment. This arises from a clearer sense of personal responsibility and a commitment to do something. It centres on individual resolution of the question, can one person make a difference? In order to feel empowered, students need to be able to envision positive scenarios for the future and to learn about success stories in which individuals and groups have clearly made a difference. There needs to be hope, humour and cautious optimism.

*Action dimension*

If the questions raised by the first four dimensions of learning have been satisfactorily resolved in some appropriate way for the student, then informed personal, social and political choices and action can occur. Some of the students Rogers interviewed reported that learning about global futures had eventually led to a significant reorientation of their lives, personally and/or professionally. Such major choices, she notes, also need to be acknowledged and supported as an outcome of the learning process.

*Figure 10.1* Learning about global futures

truly effective teaching about global issues necessitates 'three awakenings' – of the mind, the heart and the soul. Only then can a grounded personal path of action be found.

## A year's investigation

Until recently, global futures as a subject was available to first-year students within the modular scheme at Bath Spa University College. It comprised two introductory modules, which explored issues to do with land and water, energy, sustainability, human resources, alternative futures and action for change. This study explores student responses to their first year of learning global futures.

I had long felt that effective teaching about global issues was a more complex endeavour than it might at first seem. Reading Rogers' paper (described above) both confirmed this and also gave a conceptual framework for further exploration. During the first semester of the course there were always some students who reacted with concern, if not hostility, at being exposed to current issues relating to the state of the planet. At one staff–student committee meeting it was reported that the subject has already been nicknamed 'the doom and gloom course'. Since this was not one of the intended learning outcomes for the course, it seemed important to explore things in more depth.

There were three main purposes for this study: (1) to monitor in more detail the impact that learning about global futures had on students; (2) to see whether Rogers' conceptual model could be applied in an undergraduate context; and (3) to identify ways in which the research process itself might aid reflective learning.

As a result of an open invitation to the 1998–99 intake of global futures undergraduates, four students (two men and two women) agreed to take part in a year-long pilot project monitoring their responses to learning about global futures. Although we would have liked a larger group, we were entirely reliant on volunteers. The four included a mature student, one student who had come direct from school and two who had taken a year out. Data for the study were collected in three main ways. First, students kept a personal journal in which they made weekly entries in response to their global futures lectures and seminars. Second, the group met each month to share what they had written and to reflect together on the learning process. Third, at the end of each of the two modules, time was spent looking back over the unit and reviewing what had gone on individually and collectively. This is thus a qualitative ethnographic study exploring the responses of a small group of students over the course of a year.

Why did these students decide to study global futures in the first place? The subject was seen as interesting because of its focus on contemporary issues and this was a significant factor for all of the students in making their choice. One student related her choice to personal concerns about the environment and saw global futures as offering a focus for this.

> Experientially, I have seen the effects of deforestation on the environment, on indigenous populations . . . as well as the effect of Indonesian forest fires. . . . I am concerned about this. . . . Global futures seems to offer the 'right mix' to enable me to find my own answers.

Another student said that it was the description in the prospectus that caught his attention:

> It seemed like an incredibly sensible course. . . . I wanted to study something contemporary and moral, that would lead me into a life based on principles and not money.

He then added:

> I am very excited about this course, but I am also nervous. I am scared that once I know about the true situation I will get to the point where I have to do something about it. It surprises me that this scares me, it is something I should be excited about.

Students identified potential benefits for *themselves* as a key factor in deciding to join the research group. These were related to the support they hoped to gain from sharing their ideas with others and being able to develop confidence as first-year students.

> Returning to student life [after a year abroad] was very daunting to me. . . . I looked into joining the research group as an extra support, a chance to share my difficulties and get support within a small group.

> I was attracted partly through my sense of adventure. . . . I am sure I will reap the benefits of participation in any research project. How, or in what way, I don't know.

> Knowing that I have to write down how I feel about the course for someone other than myself means that I will actually get more out of it. . . . It will probably help me understand myself a little more.

At different times throughout the course each student was challenged by the complexity of the global issues they were studying, and they realised that this required them to think in greater depth and consider a greater range of causal factors than they previously had done.

> I had been aware of some or most of the issues springing from the lectures . . . but the potential to explore them further and debate them interested me greatly and I began to consider them at a deeper level.

> My awareness has definitely increased a lot. . . . It is frightening but also exciting. . . . I have never had enough facts to really worry before.

At the monthly meetings the students were able to share with each other some of what they had written in their journals. This discussion often triggered further ideas and responses. They talked of their increasing awareness and how they felt they were becoming more knowledgeable than their peers about the state of the world. They felt they were learning 'more than just academic things' – a process that was also often quite uncomfortable and even painful.

There was also a significant affective response to their growing engagement with global matters, and all of the group at some point expressed feelings of pessimism about the global issues they were beginning to explore.

> Why does it all seem so straightforward on paper yet so complex in real life? . . . This week I came away feeling a little pessimistic.

> Some of the issues are very engaging and also very sad to me. I find it difficult to stay optimistic about the future of the world . . . there seem to be precious few signs of hope.

A mature student went on to opine that the course had turned into an inward journey for her. As she wrestled with so much disturbing new information, she even wondered if it was 'maybe a mid-life crisis for me'. The students also wrote about their experiences of sharing what they felt with each other in the research group.

> More prominent this week were the inner feelings I discovered through talking with others in the group . . . it was good to learn how others were coping and how they shared similar concerns and experiences.

> I am more willing to open myself out to others – to be warmer, to be more responsive and to take the risk of being hurt in the process.

That studying global futures had an emotional impact on students was a common thread which in varying degrees ran through their journals and discussions. It was not always very overt, more a sub-theme, with a wide range of emotions being expressed at different times throughout the study. Words used to describe these feelings included: 'denial', 'frightening' and 'saddened' from one student, 'worrying', 'shocking' and 'upsetting' from another, while one spoke of being 'alarmed', 'ashamed' and 'disheartened'. There were also strong positive emotions expressed with the words 'inspired', 'enthused', 'passionate', 'excitement' and 'optimistic' used to describe the way in which they had been influenced by some of the issues being raised.

For some these feelings were very painful.

> For me, the whole module is as much a continuation of my journey inwards as it is forwards. I am discovering a lot more about myself, and more often than not it is a painful journey. . . . I am now more painfully aware of certain issues and I have to take sides and not sit on the fence. To act requires courage. . . . To be open we must drop our guard and be transparent about what we think and how we feel – initially I felt overwhelmed.

> Very personal and quite painful . . . ignorance *is* bliss . . . very depressing in a lot of ways. I can't see a way out yet . . . changing thought patterns is uncomfortable . . . feelings of futility. I find it difficult to stay optimistic about the future of the world, there seem to be precious few signs of hope.

At the end of the first semester, students were asked if they still wished to continue in the group until the end of the year. There was unanimous assent to this, because they felt they were gaining an enormous amount from the process of reflective journalling and discussion. It was seen as offering added structure to their work, and they enjoyed listening to how each other felt. One commented, 'We're not just doing it because we've got to do it, its beneficial to ourselves.'

Looking back on the experience of studying global futures and on their participation in the research group, each of the students stressed the importance of the process in gaining greater understanding of the issues and in developing deeper insights into their own responses to these. Their comments at the conclusion of the year's study included the following:

> I have learnt to be analytical and to share my feelings within the group and that I really enjoy global futures. . . . The issues that aroused the most passionate feelings were undoubtedly the ones that I could personally relate to.

> I have learnt that I am more affected by the issues than I thought I would be. Some I have found very emotionally challenging and, at times, hard to deal

with. . . . I got a tremendous amount out of the research group, being able to share how I felt and learning about how others were affected was wonderful so early in my college time.

Global futures does make you look at the photograph and ask, 'What's going on behind things?' It encourages deeper thinking . . . it has made me a more reflective person.

It added a different dimension to my life and made me more aware of my own self. . . . Sometimes it is difficult to articulate how I feel because it means entering into [the] great depths of my affectivities that sometimes tug in different and opposing directions. . . . At other times reflective journalling is a catharsis as I unburdened my thoughts and feelings and [it was] therefore therapeutic in its own way.

## Research findings

It is clear that the process of learning about global futures had a significant impact on this group of students. Whereas individual teaching staff only had a four-week slot for their particular contributions, the students had to deal with the cumulative effect of facing a whole year of global dilemmas. Issues of wealth and poverty, justice and injustice, unsustainability and sustainability are complex and can be quite overwhelming when first encountered, as the research group found. Issues of human suffering and environmental damage are uncomfortable to deal with, and it is this which adds an extra pedagogical dimension to learning about global issues.

A significant part of the impact on students was thus affective, as they began to wrestle with the implications of the case studies with which they were presented. Their responses ran the gamut from upset, worried and sad to frightened, alarmed and shocked. They also veered between pessimism and optimism, as at other times they felt inspired, enthused, excited and passionate. These findings echo some of the experiences of the Canadian students described by Rogers.

Each of the students had also expressed personal reasons for wanting to explore global futures. These related to experiences such as a year out in South Africa, travelling in Latin America, living in South-east Asia, and wanting more information on issues that 'affect our world'. The year not only gave them a wider perspective from which to evaluate their experiences but it also often problematised their previous views of this experience. The impact of the year was thus perhaps more complicated than they might have first imagined.

What evidence is there, then, for the five dimensions of learning that Rogers identified in her research? It should be noted that her model was derived from a one-year self-contained course that was probably more coherent than the experience of these students on the first year of a longer course. However, there are many similarities. Although none of the group saw the course as being 'airy fairy' (one category of response to the 'cognitive dimension' noted by Rogers), they did often feel 'cognitively overwhelmed' and periodically pessimistic about the human condition.

The 'affective dimension' was well represented as they grappled with the range of feelings that came up as part of that learning experience. Their emotions were indeed often conflicting and in particular occurred when, for whatever reason, the issues felt more personal to them. There were also glimpses of the 'existential dimension', as some of the group tried to make deeper sense of the questions they were being faced with – 'an inner journey, almost a mid-life crisis'.

The 'empowerment dimension' and the 'action dimension', however, are not really evident in this research, probably because of the nature of the first year of the course. Issues of empowerment and action were highlighted more in later second- and third-year modules. Certainly, questions about the nature of empowerment and possible choices of future action appear as a sporadic subtext in the background of students' writing and discussion.

Did the research process itself aid reflective learning? The value of guided journalling and structured discussion was noted by everyone in the group. Joining the research project gave an added focus to students' work and opportunities for individual reflection which were greatly appreciated. The monthly meetings added another dimension because this provided an opportunity for *shared* reflection and 'thinking out loud'. Some students also valued this because it helped to anchor and ground their experience as they coped with the shift into higher education.

At the end of the first semester individuals in the group were asked if they wished to continue with the research process or not. They unanimously said that despite the extra work that this involved (such as keeping a journal and attending meetings), they wanted to continue for the rest of the year. The research process thus also appeared to be responding to a felt need for reflective collaboration among the students.

In particular, the group was marked by high levels of enthusiasm for learning, journal writing, group discussion and active sharing. The fact that a regular structure was offered for this was what made this a different learning experience from any others they had in their first year. The research group was thus also a very positive learning experience and as such greatly appreciated by the students involved.

## Some conclusions

The findings from this experimental research project have to be provisional, not least because of the small number of students involved, even though they were fairly representative of the year group as a whole. As noted above, there were also differences in emphasis and structure between this particular course and the one monitored by Rogers in Canada. However, several common themes are clear. The value of Rogers' findings lies in the fact that they alert teachers to dimensions of learning often overlooked in the literature of social, environmental and global education. Because for many these dimensions are not considered to be important facets of teaching and learning they are overlooked, and their signs *not recognised* when they do appear. In part, this relates to the models of learning utilised by staff responsible for designing and teaching courses on global issues.

Learning in schools and establishments of higher education is still largely treated as a cognitive affair, with some attention possibly being paid to attitudes and values where this seems appropriate. This is part of the Enlightenment heritage in which the cognitive is valorised over the affective. It thus becomes 'natural' for learners to resist the affective, or to see it as part of the personal domain, rather than an integral element in the educational process. It was the authors' interest in more holistic models of teaching and learning which led to the setting up of this research project.

It is my contention therefore that many educators, despite their commitment to global understanding, often only make things worse for students by teaching about global issues as if this were solely a cognitive endeavour. At the very least, this and other fields relating to exploration of the human condition have cognitive, affective and existential dimensions. They also raise crucial questions about the role of education in promoting empowerment and action for change. Teachers often get it wrong when teaching about global issues because they are not equipped, personally or professionally, to deal with the affective or the existential domain. Or if they are, they are more likely to be working in a pastoral role dealing with personal and social dilemmas rather than global ones.

Some of the antidotes to this situation have, of course, been noted by those working in the field. Thus Huckle (1990) comments succinctly: 'If we are not to overwhelm pupils with the world's problems, we should teach in a spirit of optimism. We should build environmental success stories into our curriculum and develop awareness of sources of hope in a world where new and appropriate technologies now offer liberation to all.' Jones (1998), in describing his work in futures studies, writes: 'My own experience is that students get "turned on" when they are empowered and challenged to come up with strategies to cope with and adapt to the forces of social and technological change ahead.' Hicks (in Hicks and Slaughter 1998), in turn, has highlighted the personal and professional importance of identifying the sources of hope that inspire socially committed educators and others in their work. Kaza (1999) describes bringing the principles of Macy's work into her own environmental studies teaching: 'By leading people through a process of waking up to their own feelings for the world, the work releases bound energy which can then be engaged in positive effort.'

There is also a further paradox at the heart of this issue. It relates to the question, by what *right* do we expose young people to the extensive traumas of the world? Is this not some sort of betrayal of those students who do have a great enthusiasm for life? Many first-year global futures students arrive with what might be described as false hope or unrealistic optimism about the world, partly through lack of knowledge but also because they themselves have yet to be tempered by life. In a sense, the task is a mythic one – to shatter their innocence about the human condition so that the real journey can begin. The real betrayal would be *not* to awaken them to the human/global condition. As Macy and Brown (1998) argue, the heart has to be broken in order for there to be an awakening of wider compassion. A true sense of empowerment thus comes from both head *and* heart – but this requires educators who have also worked through these issues for themselves. Only then can we really begin to help students gather resources for a journey of hope.

Chapter 11

# Questioning the century
## Shared stories of past, present and future

### Summary

Periods of rapid and turbulent change affect societies in a variety of ways, including, in the 1990s, a reawakening of interest in millenarian themes. This penultimate chapter first reviews some key elements of the millennial tradition and then notes the lack of serious attention given to public perceptions of the year 2000. Second, the chapter reports on an experiential investigation in which participants explored what meanings the twentieth century, the millennium and the new century held for them. In investigating their histories and hopes they found that the personal and the political were inextricably interrelated and that the stories they shared of past, present and future led them back to the deeper perennial questions of human existence.

### The millennial tradition

> There can be no doubt; it is happening again. Another century is approaching its end; another century is about to begin. . . . And with the inevitable countdown to the new millennium has come a flood, even more copious than usual, of all those overheated fantasies of destruction and rebirth that somehow seem to attach themselves to decisive turns of the calendar page.
>
> (Jay 1993: 84)

What are we to make of that notion of the millennium – birthed in Christian cosmology yet now assimilated into popular contemporary thought? Is this a concept worthy of serious consideration or one best left to clerics, religious fundamentalists and proponents of the new age? 'Confusion about the meaning of the millennium', writes Thompson (1996: ix), 'is an enduring feature of western civilisation, and the predictable reawakening of interest in the subject during the 1990s has done little to clear it up.'

The notion of the millennium, coming as it does from the Book of Revelation, is at root a Christian concept. It could thus be dismissed as irrelevant to many of the world's population were it not for its gradual secularisation over the centuries and its absorption into Western culture which has, in turn, had an impact far beyond its European birthplace.

The last book of the Bible portrays a great battle at the end of time between Christ and Satan (or good and evil), followed by a thousand years of peace – this is literally the millennium – at the end of which comes the Last Judgement. In this context the millennium is not just the *passing* of a thousand years but an actual period of time. That this would actually occur was a powerful belief in the early Christian Church and, over the centuries, the notion of a final end to history became embedded in Western cosmology. At the end of the twentieth century we joked about PMT (pre-millennial tension), but we also look back on what historian Hobsbawm has described as the most murderous century on record (Eric Hobsbawm 1995).

The Book of Revelation is also known as the Apocalypse (from the Greek, meaning 'to unveil'), and this term is used to describe the end of the world or any titanic struggle between good and evil. Apocalyptic texts have often been Jewish but may originally have come from the Persians. They combine violent and grotesque imagery with a cataclysmic struggle between light and darkness and also glimpses of a transformed world (Thompson 1996). During the medieval period in Europe, no one doubted that the world *would* end after some such terrible conflagration between good and evil. This was an accepted fact about the social and cosmological order.

At the same time, notions of 'the end of time' often particularly appeal to groups who feel disorientated, whose identity – maybe whose very existence – is under threat. Thompson, in his book *The End of Time: Faith and Fear in the Shadow of the Millennium*, notes that

> we should not underestimate the susceptibility of the human mind to apocalyptic ideas, especially at times of rapid change. Apocalypticism, which developed as a genre at a time of acute stress for the Jews, feeds on uncertainty and disorientation . . . [and] is astonishingly versatile. End-time scenarios have the ability to adapt to their surroundings through a rapid process of mutation.
> (1996: 61)

In its narrowest sense a millenarian is someone who lives in daily anticipation of the end of the world and the subsequent dawning of a new age of peace and harmony. It is often suggested that in the year 1000 many Europeans expected the world to end. Certainly, some preachers and chroniclers saw the turbulent events of the tenth century as fulfilment of the Last Days, but in fact, historians point out, there was no standard response to the year 1000, largely because most people then had no notion of either decades or centuries (Focillon 1971; Landes 1997). Cohn (1997), in writing about medieval millenarianism, proposes as millenarian any religious movement inspired by a fantasy of salvation which is: (1) collective to the group; (2) to be realised on earth; (3) imminent – that is, due soon and suddenly; (4) will totally transform life on earth; and (5) will be accomplished by supernatural agencies.

Millenarian movements have occurred in many historical periods, the earliest being the messianic hopes of the Jews as expressed in both the Old and New Testaments. Since Christianity developed out of Judaism it inherited this expectation of a future golden age. Studies of millenarian groups in the medieval period – the adepts of the Free Spirit, the Joachimites, John Ball and his followers – show that they often seemed to occur at times when severe social and economic disruption had fractured the security of the existing social order, causing widespread public anxiety and fear (Cohn 1997).

In the seventeenth century, many religious thinkers, such as the Ranters, Diggers and Levellers, felt that they were witnessing the events leading up to the millennium. The Civil War and the execution of King Charles, unparalleled events in British history, were thought by some to be the climax of human history. Biblical prophecies, folk wisdom and astrology were intertwined to predict an immediate end-time (Popkin 1995). In the eighteenth century, apocalypse was redefined in secular terms since Enlightenment scrutiny of biblical texts had challenged literal belief in prophecy. Now, events such as the French Revolution stirred millennial enthusiasm and longings (Schaffer 1995).

The term *fin de siècle* was coined in 1885 in France to describe the social and cultural atmosphere being experienced at the end of the nineteenth century. Thompson writes:

> The notion that the new year witnesses a renewal of time is itself as old as humanity; but the idea that society grows old with the calendar century, and is mysteriously regenerated by its turn, is of far more recent origin. The sense of *fin de siècle* could not evolve until people began to feel that their lives were somehow shaped by the century in which they live.
>
> (1996: 104)

As late as the eighteenth century only official documents mentioned the AD year and most people still used the year of the king's reign. A few people may have welcomed in the eighteenth century whose end was momentously marked by the French Revolution and the Napoleonic wars. Not until 1800 were people really struck by the change of century. In the 1890s many commentators saw the approaching new century as one of bountiful social and technological innovation.

This held true until the devastation of World War One in which the dark 'shadow' side of technology was revealed in its full horror. In the face of such a prolonged dystopia how could the future remain a place of promise? The depression, the rise of fascism, World War Two, the death camps, Hiroshima, the nuclear arms race, increasing damage to the environment, the growing gap between the rich North and poor South – millions have suffered and perished in this century in apocalyptic circumstances. Apocalypse *is* at our fingertips in a way that it never was before.

At the end of the twentieth century we found ourselves facing both the fruits of modernity and its nightmare excesses. Everitt wrote:

> At the end of the last *fin de siècle*, the key to the future was held by the likes of Sherlock Holmes. The application of scientific methods to the problems of the present would allow for a benign outcome in the future. As we approach not just another end of century but the end of a millennium, poor Holmes is dead and buried. Our equivalents of the sleuth of Baker Street are Judge Dredd, Robocop, Blade Runner and Terminator, mythic figures in whom futuristic technology and medieval visions of hell have come together to form a nightmarish anti-utopia. We seem to be moving towards 2000 with a head full of fears, to be drifting into a cyberspace populated by monsters from the deep.
>
> (1995: 5)

The widespread angst of the late twentieth century and the reasons for it, from globalisation to living in high-risk societies, have been much discussed (Bailey 1988; Giddens 1991; Staub and Green 1992). It was a period ripe for stirring the millennial imagination.

The late twentieth century thus had its own share of millennial cults. These included strong fundamentalist Christian millennial traditions in, for example, South Korea and the United States; the Waco siege in Texas in 1995, where force by the authorities led to a bloody self-fulfilling prophecy; and the Aum Shinrikyo cult in Japan, acting to bring about its prophesied end-time by the release of sarin gas in public places. The New Age movement combines a rich mixture of crystals, crop circles, channelling and alien abductions with prophecies about catastrophes to come and the possible salvation of the planet through personal spiritual change.

One of the central features of the late twentieth century was a strong sense of endings, and this was present in many forms. Kermode (1968) suggested thirty years ago that human interest in endings, of whatever kind, may derive from an existential wish to halt chronicity. It may also be that our interest in endings is about the deep human need for satisfactory completion of things in our lives. More recently, Kermode noted that

> The ends of centuries and *a fortiori* of millennia are very convenient termini, either of the world or of epochs. Their attraction lies partly in their cyclical character – as we celebrate birthdays and other anniversaries – and partly in the fact that they mark or threaten a linear ending. Yet to attach grave importance to centuries and millennia you have to belong to a culture that accepts the Christian calendar as definitive, despite its incompatibility with other perfectly serviceable calendars.
>
> (1995: 251)

In writing about the nature of postmodern society, Kumar (1995b) has also drawn attention to contemporary fascination with endings and beginnings and the associated sense of hope and despondency, confidence and despair that this can bring.

The end of history, the end of science, the end of modernity, the end of communism, all have been variously pronounced. Whatever we are to make of these individually or collectively, says Kumar, we are faced at the turn of the century with the argument that the Western world is undergoing one of the most profound transformations in its history.

## Questioning the millennium

It was in the light of the above that I became interested in contemporary perceptions of millennial times. Although the millennium had been written about in popular terms (Hanna 1998), there was little serious investigation into the meanings that people ascribed to it for themselves. And any such meanings would only make sense if set within the context of the old and new centuries.

In reviewing developments in qualitative research today, Denzin and Lincoln (1994) stress that we are in a period of discovery and rediscovery in which many new ways of observing and analysing are now available. In particular, qualitative research has distanced itself from the positivist paradigm, with its belief in the possibility of objective truth. Qualitative practitioners acknowledge the impossibility of neutrality in the research endeavour, indeed in the human endeavour, and ethnographers increasingly recognise the need for 'messy' texts. Lather (1986) writes about the need for research as praxis: the quest for emancipatory knowledge which draws attention to possibilities for social transformation. Behar (1997), in her book *The Vulnerable Observer*, calls for an anthropology which involves both intellectual *and* emotional engagement and in which the researcher's openheartedness allows much more to be revealed. And Denzin argues that

> The ethnographer's tale is always allegorical – a symbolic tale that is not just a record of human experience. This tale is a means of experience for the reader. It is a vehicle for readers to discover moral truths about themselves. More deeply, the ethnographic tale is a utopian tale of self and social redemption, a tale that brings a moral compass back into the reader's . . . life. The ethnographer discovers the multiple 'truths' that operate in the social world – the stories people tell one another about the things that matter to them.
> (Denzin 1997: xiv)

This, then, is an allegorical tale about the stories we tell each other in order to make sense of our own lives and the world around us in these turbulent times.

In the light of the above, the methodologies chosen for this investigation were ones which had provided rich data on previous occasions. The focus of the research – people's perceptions of past, present and future at the millennium – required a simultaneous focus on four directions: participants' inner and outer worlds, and backward and forward in time. In order to explore these themes effectively, seventeen participants, mostly but not all working in education, volunteered to take part in a residential focus group weekend in spring 1999. This was specifically designed to encourage a participatory and cooperative inquiry into people's perceptions of the twentieth century, the millennium and the new century. A combination of individual journalling, on-the-spot transcribing, small group summaries, and experiential activities was used to explore participants' responses to the following questions:

- What does the millennium mean to you?
- What does the twentieth century mean to you?
- What does the twenty-first century mean to you?
- What are the sources of hope you draw on in your life?

## A participatory exploration

Prior to the focus group weekend, participants were asked to reflect in their journals on what meaning, if any, they gave to the millennium. Some wrote about this briefly and others at greater length. Overall there was a spectrum of opinion ranging from, on the one hand, strong rejection of the notion to, on the other, the view that this was a time of great symbolic importance and positive change.

Comments from those who rejected the notion, at least initially, included the following:

> The millennium means *nothing* to me as an event. It's just a number of mathematical significance to our so-called Christian culture.

> The millennium in itself does not mean much to me. I had been increasingly sceptical about the whole hype (the millennium dome, etc.) and was feeling 'What real difference does a year make?' Not being a Christian I was also out of sympathy with the idea that the anniversary of one religion should be regarded as more important than others.

Others were more neutral, not seeing much value in the millennium *per se*, but aware that the year 2000 might still be a useful focus.

> Whose millennium? Only a millennium for Christianity. Turning *century* feels significant and easier to think about.

> The millennium doesn't mean a lot at the moment. Is it irrelevant or do I just not want to address it?

> I have been moving towards a feeling that celebration is important and that it seems there is just some desire to celebrate together. Does it matter if we are not clear what it is that we are celebrating? Now that we are writing 00 on forms I find that it gives me a jolt. It is in some way shocking and I feel that using 2000 or 00 may have some deeper impact on us.

Several participants saw the millennium as a time of opportunity:

> A time of fundamental change; a re-appraisal opportunity; an obligation to review; a time to assert values.
>
> A chance for the world to change direction towards greater social equity and environmental protection.
>
> Celebrating a new era. Can we do better personally and globally? Time to examine our lives and our consciences.
>
> Key words: fulcrum; resurgence; turning point; renewal; spiritual rebirth.

Only the last comment really begins to echo some of the millennial themes described earlier in this chapter. Among the most reflective and far-ranging comment was the following:

> At one level the millennium is a simple way of enclosing a period of time – neat, rounded and tidy. It is a way of imposing on human experience a mathematical precision that allows us to enclose it, and control it – an arithmetical stab at describing the mystery. At another level the millennium reflects our obsession with time and our feeble efforts to control and manage it. I struggle with the difference between timeful experiencing of life, and timeless experiencing of it.
>
> Who says its the millennium anyway? In my calendar this is simply year 58. I'm sensing a strong resistance to the notion that there is particular work to be done because dates have lots of zeros on them. And yet – there is a time to look back and a time to look forward. At the weekend it will be a special time with particular work to be done. I feel ready to give myself to the experience, and find in this concentrated weekend insights and support in my struggle to find the essence in my life.

So, at the commencement of the weekend participants came with a range of perceptions about the significance of the millennium. All agreed, however, that at the very least it could be used as an opportunity to reflect on where they were in their own lives.

Participants were asked to bring with them family or other memorabilia which captured the essence of key events in the twentieth century. These were used to create a display for each decade set against a timeline of the century 1900–1999. In encouraging deeper reflection on what the passing century meant to participants, the importance of both personal family history and wider national/international events was emphasised. After sharing in small groups what they had each brought, a series of group presentations was made focusing on the highs and lows of each decade.

This experiential exploration of the century was a powerful one. Whether we were young or old, this was the century we had all been born into and in which our parents and grandparents had lived. The collective sharing often resulted in deeper insight into how the personal and public are always intertwined.

> My feelings are quite painful as I talk about the reasons I am here – I have this picture of my grandmother at the age of 12 arriving in England from Germany having fled her home. She had travelled with her mother the last part of the journey from Russia. The village she lived in with her family just outside Kiev had been burnt down in the pogroms – and ethnic cleansing is still with us – look at Kosovo.

One participant who was born at the end of World War Two read from her mother's journal for that period:

> *23rd August 1944* – The Marseillaise thrilling the world, Paris liberated by her own citizens. My bonny baby hears it, not knowing in what stirring days her life begins. Pray God they may never come again for her hereafter! *24th December 1944* – So ends a year which has sent my husband to the battlefield and driven me as a refugee to other people's houses, yet which has brought me the greatest of joy in my baby, if I can keep her.

Reflecting on the stories he had heard, one person commented:

> How different family stories are to tell and hear when they are not simply recounting – but digging for the meaning. How we carry the experiences so vividly all these years after. I struggled with the relationship between being a witness of this century and being a participant.

Another commented:

> An incredible period of revolutionary change in the planet's history. The same old human frailties and obsessions have been played out again and again but with amplified consequences due to increased technology, globalization and population increase. Also a period when the great narratives have been exercised and found wanting, e.g. fascism, communism, totalitarianism, etc.

Most people found this shared exploration of the century a powerful one, summed up in the comment, 'Touching base with the whole century was amazing through other people's stories and portrayals, the personal with the political.' It was seen as a century of great change and progress but also of great pain and trauma.

Having been present to the past by looking reflectively at the century they were born into, participants were now faced with being present to the present. Having just reviewed a tumultuous and often traumatic century and with an unknown future waiting for attention, the present felt a little uncomfortable. Participants spoke of crossroads and decisions in their own lives, of a sense of waiting and transition, of feeling clear and confused at the same time, of a huge change of consciousness during the century, and feelings of privilege and responsibility at being in this space.

> Discussing the present has made me quite uneasy. I have far too many questions and uncertainties. I'd rather be in denial – putting off these decisions for as long as I can get away with it. I'm looking forward to discussing the future – perhaps some favourable outcomes may come to light during the session.

> The group presentations didn't really work because each person is facing their individual dilemmas and it didn't/couldn't link into any *fin de siècle* Zeitgeist. I want to get into the next century now.

> I still can't look forward to the millennium, in either the practical or optimistic sense, until I have dealt with my present fears. Why do we foolishly believe that a change in the date will alter our lives unless we are first not capable of altering ourselves?

Several participants were in their fifties and found that they, and others, were faced with their own mortality in the present. 'What does living really mean knowing that I'm going to die?' and 'How *do* we give significance to our lives?' were two of the existential questions that arose. There was thus some apprehension and uncertainty, both personal and societal, in this space, a sense of transitions, of being at the crossroads, of having to face multiple decisions.

One of the tasks prior to the focus group had been for individuals to reflect on what the new century meant to them. For one person it was about postmodernity, instability, globalisation and insecurity. Another saw it in more personal terms: the second half of my life, being there for my children and grandchildren, getting old, dying, but also 'sustainability as the measure of all things'. Some had questions: 'How much of it will I experience? How much of it will my son experience and what will it be like for him? What will I miss out on that would have enthralled me?'

> I'm hesitant about attaching any meaning to the new century as I'm sceptical about the usefulness of packaging eras in this way and labelling them. There is a strange irrational feeling that somehow all of us are being set adrift in a new world without security, as though it was really all a new page. I think this is exacerbated by the millennium bug business.

> 'Necessity is the mother of invention', so I am optimistic that the current threats and challenges will actually galvanise the human race into re-evaluating past history and exploring new, more appropriate, ways of being. Much as the quality of life for a patient with a life-threatening illness can be enhanced through reflection and re-evaluation. We may not find the cure in time but the process of looking will be a really healthy one.

*Table 11.1* Preferred futures, 2050

*Atmosphere* – a feeling of freedom and safety; no fear or worry; slower pace of life; conviviality; gentle noises; a spiritual theme; no consumer goods; a sense of collective responsibility; only the hum of laughing, talking, shouting

*People* – lots of greeting and communicating; children everywhere asking questions, having fun; people know each other; multicultural; all ages; community groups; no work/play separation; cooperative; peaceful, warm, serene

*Environment* – light, greenery, things growing; a focal space; sunshine; water, fountains, gardens; glittering meadows; trees and woodland; birds, flowers, wildlife; clean; bright colours; low-tech buildings; houses with shrines; cafés, markets, small shops; computers; open schools

*Technology* – alternative energy; water power; wind generators; solar power; craft workshops; recycling centre; reedbed sewage system; human technology blending into the landscape

*Food* – luxuriant vegetables growing; fruit bushes; vineyards; local food; fresh food

*Transport* – no traffic noise; no need to travel far; cycles, walking; no cars; people flying; bikes, roller blades; light railway

In the latter part of the focus group participants returned to explore further their feelings about the new century, this time by investigating the nature of their preferred futures for 2050. After using the visualisation procedure developed by Boulding (1994), participants worked in small groups to create posters illustrating the main features of their desirable futures. A composite list of the descriptors which arose is given in Table 11.1 above.

It should be noted that there were some significant differences within the group arising from the visualisation. On commencing this envisioning some people found themselves in places that they knew in 1999, while others were in places not known to them. Most found themselves in the countryside but some were in cities. Although most felt in a safe and secure place, a few participants had a sense that something unpleasant had occurred between 1999 and 2050, either in their area or elsewhere.

Reflecting on their joint visions of 2050, participants were particularly struck by: the human scale of society, powerful new forms of intimacy that existed, the different quality of life at a slower pace, the localised nature of the economy, the medieval quality of their shared community. 'But if this is the Garden of Eden', someone asked, 'where is the serpent?'

To work towards their desired futures, whether in a personal or professional capacity, people also need to identify the sources of support and inspiration that will help motivate their endeavours. Each participant had therefore been asked to bring an object that for them represented a source of hope. Sitting in the final closing circle, people took it in turns to say what their symbol was and then to lay it with the other objects on a table in the centre. The different sorts of sources that participants draw on are shown in Table 11.2.

*Table 11.2* Sources of hope (3)

*Nature* – a cup of water from the spring; a stone to start a building; planting a garden; pebbles from a beach; herbs from a garden
*Creativity* – reading and poetry; a collectively made quilt; the internet; a Chinese wall scroll; a hammered dulcimer
*People* – friends and family; a quotation from Margaret Mead; photos of a class of children and also of a group of students
*Time* – the gifts from the century identified by the group earlier; future times with family
*Life* – accepting change as ever present in our lives; always being open to inquiry

## Reflections on the experience

Any analysis made from these data, any picture of the shared exploration described here, will be partial and only one account out of many possible ones. This version was sent to all participants for their comments and suggested amendments. Despite its partiality, I believe it does capture the spirit of the focus group and much of the reflection and interaction that took place within it. What has been learned, then, about the responses of this group to the four research questions?

The notion of the millennium having any spiritual or secular significance was rejected by most participants. For most, the millennium meant: (1) the hype associated with the Millennium Dome being built at Greenwich, or (2) just another date on the Christian calendar. Some, however, did have a sense that this turning of the century might be put to good use. Only one participant spoke of a sense of deeper spiritual and socio-cultural change being afoot. All acknowledged that it could provide an opportunity to spend time reflecting on where they were and where they wanted to go in their own lives.

Despite its man-made nature, the ending of this century was seen as a useful point from which to assess the last hundred years. The fact that everyone, to a greater or lesser extent, could chart their family histories through this period made it both an exciting and disturbing exploration. It was the doing of this as a shared and collective act that made it so powerful. Listening to the stories of other people's parents and grandparents brought the decades painfully alive and reminded us of our forebears' fortitude and courage. The personal and the political could no longer be seen as separate, and the impact of history on our own inner and outer lives was made painfully clear. I think there was a sense of awe that people could have lived through such times and somehow survived with a measure of dignity.

When faced with their own present and the collective present the group was distinctly uncomfortable. Some people were able to speak about their discomfort and uncertainty; others repressed it or did not feel it. What is certain is that the present will inevitably feel daunting if consciously experienced in the light of all that has gone before. This applies, whether one is thinking just about one's own life or wider changes in global society. To separate the present from the past may be more comfortable but, in a sense, time past, present and future are an indivisible whole. The uncertainty felt in the group at this time merely echoed, I believe, the wider and very real postmodern angst of the late twentieth century.

The future was seen as a place of both hope and fear. For older members of the group its contemplation raised existential anxieties about what was still to come or what they might not live to see. My decision to focus on preferable futures was a deliberate one because, in pedagogical and existential terms, this is something that always enables people to move on both in head and heart. Despite individual differences, the collective future visualised was a familiar one. It was the 'baseline' future that Boulding describes finding in so many of her futures workshops (Chapter 6) and which has also been identified in previous research.

Such decentralised ecotopias seem to have come nearer to the surface of the Western collective unconscious during the last thirty years. Like all utopias, they are both a critique of present industrial-consumerist practice and a vision of what an alternative good society might look and feel like. Such preferable futures are utopian in the sense that they are all-embracing in their view of society but they also partake of some elements of the millennial tradition. Thus, whereas the group generally eschewed any deeper notion of the millennium, their collective vision has some echoes of a Garden of Eden, at least in 2050 if not for a thousand years.

The final sharing of people's sources of hope was a powerful act and, if done deliberately and with intent, always shifts group energy into an empowered mode. Although less time was spent on this than on some previous occasions, the same categories of hope were again apparent (Chapter 7). It was a reminder again to me as a 'vulnerable researcher' that the overall structure of such focus groups must not only offer opportunity for critical reflection but also the opportunity in these troubled times of being able to move forward with hope.

It could be argued that this was a self-selecting group and that the perceptions of past, present and future found here merely reflect Western middle-class perspectives. However, the collective observations of the group are not ones from which any wider generalisations can be derived. They are, nevertheless, still of great interest and add something to the existing data on images of the future, not least the broader socio-historical context in which such explorations need to be set. In the end, by choosing to engage in this exploration, the participants found themselves faced with some of the perennial questions of human existence.

- Is the struggle to be human really any different today from that which our forebears went through?
- What can we learn from the struggles of past generations that will help us here in the present?
- How do we continue to maintain hope in the face of human intransigency and weakness?
- How do we stay present to the cycle of our lives and all the joys and pains that come with being human?

Days, months, years, centuries are but a temporal grid that we lay over our lives in order to try and make sense of time. The notion of the millennium merely derives from one Western attempt to do this. It is truly a socio-temporal construct possessing only the meaning that people give to it. Despite or because of this, it has been a powerful organising concept down the centuries. Neither should this be dismissed lightly in the face of millennial cults which wish to bring about their own self-fulfilling apocalyptic prophecies (Cockburn 1995).

In sharing the history of their families in the twentieth century, participants were very present to apocalypse although they chose not to dwell on it for long. Apocalypse has become something that we have learned to live with, not least after a forty-year nuclear arms race between the two superpowers. The discomfort of being present to the present was also, I believe, about a sense of burden: how can one staunch the wounds of this century in any other way than through time?

Griffin writes about the embodiment which the group experienced:

> Nowhere is there a record of all that has happened in human history, except in living consciousness. And does the truth each of us knows die along with us unless we speak it? This we cannot know. Only we know that the consequences of every act continue and themselves cause other consequences until a later generation will accept the circumstances created of these acts as inevitable. Unless this generation tries to unravel the mystery. And if they penetrate the secret whose scent persists in all eventualities, will they say, finally, this death, this wound, this suffering, was not necessary?
>
> (S. Griffin 1994: 69)

A desire for utopia, an end to human ills and the creation of a just and equitable society, always lies just below the surface of the human imagination. It is a mythic need and one that we can only keep returning to. The utopian imagination partakes of the millennial tradition in that it deeply desires a long-term resolution to issues of human conflict and pain (Kumar 1995b). The current interest in envisioning preferable futures, whether for communities, organisations, groups or individuals represents, I believe, an often unconscious welling-up of that desire. Some would say it is not surprising that this interest flourishes at the start of a new century.

The shared reflections on past, present and future described here are indeed part of the stories that we tell each other about things that matter. They are important because, as Freire (1998: 45) points out, 'without a vision for tomorrow, hope is impossible'. In volunteering to explore these issues together at the turn of the century, the participants in this group, whether they intended it or not, found themselves contributing to what Denzin (1997: xiv) has called the 'utopian tale of self and social redemption'.

# Chapter 12

# Epilogue
## Some lessons for the future

### Summary

This final chapter reconsiders the purposes of this book and the futures-orientated themes that it has explored. It identifies further research that is needed in relation to young people and the future, and reiterates the need for a futures dimension in the school curriculum. By way of example it looks at the work that lead to official UK support for a global dimension in the curriculum. Important similarities and differences are noted with regard to a futures dimension and some preliminary steps towards its inclusion in the curriculum are sketched out.

### Purposes of the book

The main purpose of this book has been to draw attention to the need for more critical and creative thinking about the future during the early years of the twenty-first century. In particular, I am interested in what this means for young people in school, students training to be teachers, teachers in schools and lecturers in higher education. I am also very aware that, in arguing for a futures dimension in the curriculum, I often initially meet incomprehension.

Dator (addressing readers of *American Behavioral Scientist*) explains why this is the case:

> the chances are very good that . . . you have never taken a course in futures studies; never met a person who teaches it at the university level; teach or study on a campus where futures studies is not offered; and probably associate *futures studies* (if the term means anything to you at all) either with astrology and charlatans or with Alvin Toffler, John Naisbitt, or Faith Popcorn. . . . Your own reading about the future is, in all probability, restricted to *Brave New World* and *1984* (if you are of a certain age cohort) and/or to varieties of science fiction and comic books. Your most fundamental images of the future are almost certainly shaped primarily by films and videos you have seen.
>
> (Dator 1998: 298)

I have to say, that after a lecture or workshop on the *nature* of futures education, nearly all teachers move from (probably) the above position to one of 'This feels really important, tell me more!'

In his foreword to this book Wendell Bell admirably sketches some of the key concerns of futures studies and the crucial importance of this academic field. I find it fascinating how many authors and reports pontificate about the future without ever drawing on the insights of futures studies. In doing so they often produce seriously flawed documents which are, quite wrongly, taken by readers as authoritative statements about futures issues. One of the purposes of this book has been to bring awareness of the value of futures studies and futures education to a wider audience. In the busy world of teachers it is more appropriate to talk about the need for a *futures dimension* in the curriculum rather than using the term 'futures education', which is not in common usage in schools.

The research described in this book is work in progress. It is incomplete but nevertheless adds considerably to the knowledge base of futures education. It deals with seven main areas of concern, which can be summarised as follows:

1. *Educational rationale*: it is important to be able to give a clear educational justification for a futures dimension in the school curriculum. The main elements of such a rationale are set out in Chapter 2.
2. *Young people's concerns*: it is important to know what young people's hopes and fears are for the future, whether their own, for their community or the world. Some of what we know about this is set out in Chapter 3.
3. *Practical application*: it is important to see how futures ideas can be applied to different curriculum subjects in immediate and useful ways. Geography and fieldwork are two contexts in which this has been done, as described in chapters 4 and 9.
4. *Envisioning the future*: it is important that students and teachers are able to envision preferred futures for the local/global community. This can lead to more effective action for change. Some key elements of such visions are reported on in chapters 5, 6 and 8.
5. *Global issues*: it is important that young people learn about the possible impact of current local/global issues on the future. Teaching about such issues, however, requires a holistic approach to learning, as reported in Chapter 10.
6. *Sources of hope*: it is important that students and teachers are aware of the sources of hope and inspiration that they can draw on in turbulent times. Some examples of these sources, and how to identify them, are given in chapters 7 and 8.
7. *Changing times*: it is important that educators help their students reflect critically on the changing times in which they live and how they can engage creatively with change. An example of educators doing this for themselves is set out in Chapter 11.

I chose to open this book with an autobiographical chapter in order to show some of the ways in which educators come to take a socially critical stance and to challenge the traditional orthodoxy that 'voice' has no place in academic work. As the chapter shows, the personal and professional are inextricably interrelated and, for me, cannot help being present in my work. My academic persona can never be separate from the personal/ professional journey that I have undergone to arrive in this place. This is why I find recent developments in qualitative research invigorating and challenging.

## Some research issues

The work described in this book is part of an ongoing research project on 'Images of the future in postmodern times'. Denzin (1997: xiii), in writing about ethnography in the twenty-first century, argues that 'the writer can no longer presume to be able to present an objective, non-contested account of the other's experiences'. I believe this to be the case, both in relation to my own involvement in education and the research projects reported in these pages. Denzin (1997: xv) further argues that, 'although the field of qualitative research is defined by constant breaks and ruptures, there is a shifting centre to the project: the avowed humanistic commitment to study the social world from the perspective of the interacting individual'. This sits well, I think, with Wagar's (1992: 34) comment that 'the future is a very murky place' with 'no eye-witness accounts, no first-hand evidence' and its exploration is always a normative endeavour.

The field of futures education is still under-researched, and there are many crucial issues and themes awaiting investigation, as shown in the box below.

---

RESEARCH NEEDED IN FUTURES EDUCATION

*Images of the future*

- How do children conceptualise time and the future and how does this vary with age?
- How do children's views of the future vary by gender?
- How do children's views of the future vary by social class?
- How do children's views of the future vary by ethnic group?
- What is the nature of children's probable and preferred futures?
- What emerges from cross-cultural comparisons of the above?

NB: Views of the future could be broken down into personal, local, national and global. The same questions could also be asked of adults and in relation to the same variables.

*Media influences on images*

- What images of the future are conveyed by children's comics, books and computer games?
- What images of the future are conveyed by TV advertising?
- What images of the future have been conveyed by popular films over the last twenty-five years?
- How do such images relate to issues of gender, age, class and Western culture?

*Image and action*

- How do images of the future affect attitudes and behaviour in the present?
- What determines reactive or proactive stances in relation to the future?
- What changes in attitude and behaviour arise from extended futures-orientated work in a school or classroom?
- What do teaching materials that encourage skills of participation and responsible action look like for different age groups?

*Resources and policy*

- What do appropriate teaching materials look like for different subject areas and how can subject specialists be encouraged to develop them?
- Which futures methodologies are most useful in the classroom and how can they be related to a range of other learning outcomes?
- How can head teachers, school governors and parents be persuaded of the need for a futures dimension in the curriculum?
- What educational bodies and which key players would need to be influenced in order to gain official backing for such a dimension in the curriculum?

This is in no way intended to be a complete list but it highlights a range of initial research possibilities. In discussing the need for a futures dimension in the curriculum, it is instructive to consider how a global dimension came to gain official recognition in the United Kingdom.

## A global dimension

Debate about the need for a global dimension in the curriculum was initially prompted by those working in global education, world studies and development education. This arose because it was seen that, on the spatial dimension of the curriculum (local/national/global), the global element was often missing in schools. Put at its simplest, the term 'global education' describes a form of education which *promotes the knowledge, skills and understanding needed to live responsibly in a multicultural society and an interdependent world*. Elaborated in more detail (Development Education Association 1999), it;

- enables people to understand *the links* between their own lives and those of people throughout the world;
- increases understanding of the economic, cultural, political and environmental influences which shape our lives;
- develops the *skills, attitudes and values* which enable people to work together to bring about change and take control of their own lives; and
- works towards achieving *a more just and sustainable world* in which power and resources are more equitably shared.

The key organising concept within global education is that of *interdependence*, which highlights the many interconnections existing between people, places, issues and events in the world today. Exploration of *local–global* connections is at the heart of global education and, in different ways, they are relevant to all curriculum subjects. Global education places particular emphasis on process as well as content, and is accordingly characterised by approaches to teaching and learning which are experiential and participatory.

These educational concerns have a long history within the United Kingdom. In the early and middle years of the twentieth century the phrase 'education for international understanding' was used by educators. Organisations such as the Council for Education in World Citizenship played a major part in establishing and maintaining this tradition. In the 1950s and 1960s Jim Henderson, at the University of London Institute of Education, emphasised the need for a world-centred perspective in the curriculum, and it was he who coined the label 'world studies' to describe this.

In 1972 Henderson helped set up the World Studies Project, through his involvement with the One World Trust. It was this project, under the influential leadership of Robin Richardson, which firmly placed world studies on the educational map. Its successor, the World Studies 8–13 project, worked with more than half the LEAs in England and Wales during the 1980s and 1990s. At the same time, the Centre for Global Education in York also developed wide-ranging networks through its in-service work and publications. During the 1990s the term 'world studies' gradually gave way to 'global education' or, more practically, the need for a global dimension in the curriculum.

From the late 1960s onwards many development agencies, such as Oxfam and Christian Aid, with their particular expertise in issues of global inequality and injustice, began to take a specific interest in education. During the late 1970s and early 1980s a number of Development Education Centres (DECs) were set up to work with teachers, and several of the larger centres developed major publishing programmes. Today, the Development Education Association acts to create a forum for debate on development education theory and practice in Britain and overseas.

The need for a global dimension in the curriculum has thus largely been promoted by proponents of global education (based in higher education) and development education (DECs and NGOs), but with important contributions from geographers and those involved in multicultural and anti-racist education. Official recognition of this need is now enshrined in a booklet for schools entitled *Developing a Global Dimension in the School Curriculum* (DfEE 2000). In a very real sense it stands on the shoulders of those who worked tirelessly over the last thirty years for just such recognition.

## A futures dimension

Can the above brief history shed any light on how best to argue the need for a futures dimension in the curriculum? While clearly there are lessons to be learned, there are also many differences. The spatial dimension of the curriculum has long been the provenance of geographers, with their emphasis on place and space at local, national and global scales. Both global and development educators have argued, however, that *all* subjects can contribute to a global dimension. The spatial dimension is a relatively concrete one, as it is about actually existing connections between local and global communities.

Whilst the temporal dimension (past/present/future) forms the other main axis of the curriculum, it is very different in nature. It is only time past which has been claimed by subject specialists in schools. Historians are quite clear about their contribution to exploring the temporal dimension. And all teachers, if asked, say that they are concerned about helping students understand the present. But the future, which does not yet exist, is largely absent from the curriculum. As the 'global' was identified in the 1970s as a missing element in the curriculum, so the 'future' was identified as a missing element in the 1990s. But the temporal dimension is more elusive than the spatial. The past has gone (although we may learn from it), the future (which we will also learn from) has yet to come. All we have immediate experience of is the present. The temporal dimension in a sense fades away in both directions although, of course, historians have artefacts and documents from which different pictures of the past may be reconstructed.

Cornish writes about the 'three paradoxes of time':

*The paradox of the future*: The future does not exist and never has existed, yet it is our most precious possession because it is all we have left. The future is where we will spend the rest of our lives. Since the future does not exist, it cannot be examined or measured or subjected to scientific tests in a laboratory; it can only be studied by means of ideas based on knowledge from the past.

*The paradox of the past*: The past is the source of all our knowledge, including our knowledge of the future. But, despite everything that we know about it and even our personal experience with it, we are powerless to improve the past or change it in any way because, by definition, the past no longer exists.

*The paradox of the present*: The present is the only period of time that exists and in which we can think and act, yet it is merely the boundary between the past and the future without any duration or existence of its own.

(Cornish 2001: 32)

These paradoxes, argues Cornish, lie at the heart of human existence, and it is partly this which makes a futures dimension so different from its global counterpart.

Put at its simplest, the term 'futures education' describes a form of education which *promotes the knowledge, skills and understanding that are needed in order to think more critically and creatively about the future*. Elaborated in more detail, it

- enables pupils to understand the *links* between their own lives in the present and those of others in the past and future;
- increases understanding of the *social, political and cultural influences* which shape people's perceptions of personal, local and global futures;
- develops the *skills, attitudes and values* which encourage foresight and enable pupils to identify probable and preferable futures; and
- works towards achieving a *more just and sustainable future* in which the welfare of both people and planet is paramount.

Some of the elements of such an approach have been the focus of this book. But what can be said about efforts to achieve official recognition of the need for such a dimension in the curriculum?

Because a detailed history of futures education has yet to be written, what follows is an impressionistic sketch of a few important elements. One of the first writers to draw attention to work in schools was Toffler (1974) in his still very relevant *Learning for Tomorrow: The Role of the Future in Education*. His key thesis remains as true now as then: 'all education springs from images of the future and all education creates images of the future'. A few years later the US National Council for Social Studies produced its influential bulletin *Futures Unlimited: Teaching about Worlds to Come* (Fitch and Svengalis 1979). A sprinkling of classroom materials, largely American (for instance, Riley 1989), gradually became available for teachers in the 1970s and 1980s.

The World Futures Society (WFS) and the World Futures Studies Federation (WFSF) were set up in 1966 and 1973 respectively, and are two of the major organisations that, among other roles, service futures studies. However, neither has spawned offshoots that specifically exist in order to service schools and teachers.

By contrast, global education, development education, peace education and environmental education arose in part because educators with an interest in the related academic fields wanted to know how to teach about these issues in schools.

During the 1980s, futures issues were taken up in the United Kingdom by some global educators, such as the World Studies 8–13 project (Hicks 1990) and the Centre for Global Education (Pike and Selby 1988). Chapter 3 of this book has reported on some of the international research relating to young people's images of the future. This research expanded in the 1990s, a decade which also saw futures-orientated initiatives in Australia (O'Rourke 1994) and innovative projects such as Creating Preferred Futures in the United States. International good practice, however, is still scattered, and it is interesting to note that those who take an interest in futures education often have a prior commitment to fields such as social education, global education and development education.

These notes can only be indicative because, unlike those arguing the need for a global dimension, those interested in futures education have yet to achieve a coherent voice. There is no clear body of opinion in mainstream education that understands, let alone supports, the need for a futures dimension. It is still the domain of a loose international network of socially committed educators. Although there are some official initiatives (Education Queensland 2000), they are few in number and may still only reproduce Gough's (1990) categories of 'tacit, token and taken-for-granted' futures.

Steps that would need to be taken in the future would include the following:

- the creation internationally of a seed group of educators specifically committed to futures education, possibly as an offshoot of WFSF;
- the creation nationally of networks of teachers and teacher educators committed to futures education;
- the setting up of alliances, nationally and internationally, with colleagues working in social education, global education, development education;
- the development of teaching materials for different age groups and subject areas that embody the principles of futures education;
- work with professional groups (e.g., teachers, head teachers, subject specialists) to incorporate a futures dimension in policy documents;
- offers of professional development programmes to schools, local authorities, curriculum development bodies and other national bodies;
- identification of key players in education who need to be inducted into the principles and practice of futures education; and
- circulation nationally and internationally, via conferences and newsletters, of examples of successful practice at all levels of education.

If official recognition of a global dimension in the United Kingdom is anything to go by, achieving the same for a futures dimension could take twenty-five or thirty years. Is there enough commitment internationally to attempt this? Who are the key players and where are they now? Who would be natural allies? What and where are the pressure points that need to be worked on? And how can the wider field of futures studies support those working in schools and teacher education to begin such a programme?

In discussing William Morris's great utopian novel, *News from Nowhere*, Coleman and O'Sullivan write:

> Let us imagine that life is not as it is, but as it one day might be. Let us inspect the unknown terrain of the future, as if we are about to inhabit it . . . the imagined future is a subversive force: the more who imagine a different kind of future, and imagine constructively, materially and determinedly, the more dangerous utopian dreams become. They grow from dreams to aims.
> (Coleman and O'Sullivan 1990: 10)

Morris, I suspect, would be delighted to know that in the early twenty-first century these concerns are still alive and at the heart of the academic field of futures studies, and futures education in schools.

# Bibliography

Bailey, J. (1988) *Pessimism*, London: Routledge.
Ballard, J.G. (1994) *Myths of the Near Future*, London: Vintage Books.
Bardwell, L. (1991) 'Success stories: imagery by example', *Journal of Environmental Education*, 23: 5–10.
Beare, H. and Slaughter, R. (1993) *Education for the Twenty-first Century*, London: Routledge.
Beck, U. (1998) 'Politics of risk society', in J. Franklin (ed.) *The Politics of Risk Society*, Cambridge: Polity Press.
Behar, R. (1997) *The Vulnerable Observer: Anthropology that Breaks Your Heart*, Boston: Beacon Press.
Bell, D. (1967) *Toward the Year 2000*, Boston: Houghton Mifflin.
—— (1973) *The Coming of Post-industrial Society*, New York: Basic Books.
Bell, W. (1997) *Foundations of Futures Studies*, 2 vols, New Brunswick, NJ: Transaction Publishers.
Bell, W. and Mau, J. (1971) *The Sociology of the Future*, New York: Russell Sage Foundation.
Berry, T. (1990) *The Dream of the Earth*, San Francisco: Sierra Club.
Bloch, E. (1986) *The Principle of Hope*, Oxford: Blackwell.
Boulding, E. (1979) 'Remembering the future: reflections on the work of Fred Polak', *Alternative Futures: The Journal of Utopian Studies*, Fall.
—— (1988) *Building a Global Civic Culture: Education for an Interdependent World*, New York: Teachers College Press.
—— (1994) *'Image and action in peace building'*, chap. 5 in D. Hicks (ed.) *Preparing for the Future: Notes and Queries for Concerned Educators*, London: Adamantine Press.
Bowers, C.A. (1993) *Education, Cultural Myths, and the Ecological Crisis*, Albany, NY: State University of New York Press.
Boyle, G. (1996) *Renewable Energy: Power for a Sustainable Future*, Oxford: Oxford University Press.
Brown, H. (1954) *The Challenge of Man's Future*, New York: Viking.
Brown, L. and Starke, L. (1996) *State of the World 1996*, London: Earthscan.
Brown, L., Flavin, C. and Postel, S. (2001) *State of the World 1999*, London: Earthscan.
Brown, L. et al. (2001b) *Vital Signs 2001–2002: The Trends that are Shaping Our Future*, London: Eartscan.
Brown, M. (1984) 'Young people and the future', *Educational Review*, 36: 303–15.

Burgess, J., Limb, M. and Harrison, C.M. (1988) 'Exploring environmental values through the medium of small groups', *Environment and Planning A*, 20: 309–26 and 457–76.
Button, J. (1995) *The Radicalism Handbook: A Complete Guide to the Radical Movement in the Twentieth Century*, London: Cassell.
Callenbach, E. (1981) *Ecotopia Emerging*, Berkeley, CA: Banyan Tree Books.
Cantrell, D. (1993) 'Alternative paradigms in environmental education research: the interpretive perspective', in R. Mrazek (ed.) *Alternative Paradigms in Environmental Education Research*, Troy, OH: North American Association for Environmental Education.
Cantril, H. (1965) *The Pattern of Human Concerns*, New Brunswick, NJ: Rutgers University Press.
Capra, F. (1983) *The Turning Point: Science, Society and the Rising Culture*, London: Flamingo/Fontana.
Carter, L., Mische, A. and Schwarz, D. (eds) (1993) *Aspects of Hope*, New York: ICIS Center for a Science of Hope.
Centre for Alternative Technology (1995) *The CAT Story: Crazy Idealists!* Machynlleth: Centre for Alternative Technology.
Cheney, J. (1989) 'Postmodern environmental ethics: ethics as bioregional narrative', *Environmental Ethics*, 11: 117–34.
Clarke, I.F. (1992) 'Twentieth century future-think: all our yesterdays', *Futures*, 24: 251–60.
Cockburn, C. (1995) 'Crossfire', *The Independent Magazine*, 22 May.
Cohn, N. (1997) 'Medieval millenarianism', chap. 3 in C. Strozier and M. Flynn, *The Year 2000: Essays on the End*, New York: New York University Press.
Coleman, S. and O'Sullivan, P. (1990) *William Morris and News from Nowhere: A Vision for Our Time*, Bideford: Green Books.
Connell, S., Fien, J., Lee, J., Sykes, H. and Yencken, D. (1999) ' "If it doesn't directly affect you, you don't think about it": a qualitative study of young people's environmental attitudes in two Australian cities', *Environmental Education Research*, 5: 95–113.
Cornish, E. (2001) 'Three paradoxes of time', *The Futurist*, July–August: 32.
Cottle, T. and Klineberg, S. (1974) *The Present of Things Future*, New York: The Free Press.
Danziger, K. (1963) 'Ideology and utopia in South Africa: a methodological contribution to the sociology of knowledge', *British Journal of Sociology*, 14: 59–76.
Dator, J. (1993) 'From future workshops to envisioning alternative futures', *Futures Research Quarterly*, 9: 108–12.
—— (1996a) 'Futures studies as applied knowledge', chap. 4 in R. Slaughter (ed.) *New Thinking for a New Millennium*, London: Routledge.
—— (1996b) 'Forward', in R. Slaughter (ed.) (1996) *The Knowledge Base of Futures Studies*, Melbourne: Futures Study Centre.
—— (1998) 'The future lies behind! Thirty years of teaching futures studies', *American Behavioral Scientist*, 42: 298–319.
Denzin, N. (1997) *Interpretive Ethnography: Ethnographic Practices for the 21st Century*, London: Sage Publications.
Denzin, N. and Lincoln, Y. (eds) (1994) *Handbook of Qualitative Research*, London: Sage Publications.
Department for Education and Employment (2000) *Developing a Global Dimension in the School Curriculum*, London: Department for Education and Employment.
Desroche, H. (1979) *The Sociology of Hope*, London: Routledge & Kegan Paul.
Development Education Association (1999) *Annual Report of the Development Education Association 1999*, London: Development Education Association.

Diamond, I. and Orenstein, G. (1990) *Reweaving the World: The Emergence of Ecofeminism*, San Francisco: Sierra Club.
Dobson, S. (1995) *Green Political Thought*, London: Routledge.
Dodds, F. (ed.) (2001) *Earth Summit 2002: A New Deal*, London: Earthscan.
Douthwaite, R. (1996) *Short Circuit: Strengthening Local Economies – Security in an Unstable World*, Dartington: Resurgence.
Dudley, N. (1991) *The Soil Association Handbook: A Consumer Guide to Food, Health and the Environment*, London: Macdonald Optima.
Eckersley, R. (1994) 'A machine at the heart of the world: youth and the future', paper presented at the forum Shaping Schools' Futures, Melbourne.
—— (1999) 'Dreams and expectations: young people's expected and preferred futures and their significance for education', *Futures*, 31: 73–90.
Education Queensland (2000) *2010: A Future Strategy*, Brisbane: Education Queensland.
Ehrlich, H. and Ehrlich, C. (eds) (1979) *Reinventing Anarchy: What are Anarchists Thinking These Days?* London: Routledge.
Elgin, D. (1991) 'Creating a sustainable future', *ReVision*, 14: 77–9.
Ellis, C. and Bochner, A.P. (2000) 'Autoethnography, personal narrative, reflexivity: researcher as subject', chap. 28 in N.K. Denzin and Y.S. Lincoln (eds) *Handbook of Qualitative Research*, 2nd edn, Thousand Oaks, CA: Sage Publications.
Everitt, A. (1995) 'The Dredd of 2000 AD', *Guardian*, 7 January.
Fien, J. and Gerber, R. (eds) (1988) *Teaching Geography for a Better World*, Edinburgh: Oliver & Boyd.
Fisher, S. and Hicks, D. (1985) *World Studies 8–13: A Teacher's Handbook*, Edinburgh: Oliver & Boyd.
Fitch, R. and Svengalis, C. (1979) *Futures Unlimited: Teaching about Worlds to Come*, Bulletin 59, Washington, DC: National Council for the Social Studies.
Focillon, H. (1971) *The Year 1000*, New York: Harper & Row.
Fountain, S. (1995) *Education for Development: A Teacher's Resource for Global Learning*, Sevenoaks: Hodder & Stoughton.
Fox, M. (1994) *The Reinvention of Work: A New Vision of Livelihood for Our Time*, San Francisco: HarperSanFrancisco.
Freire, P. (1994) *Pedagogy of Hope*, New York: Continuum.
—— (1998) *A Pedagogy of the Heart*, New York: Continuum.
Gabor, D. (1964) *Inventing the Future*, New York: Alfred Knopf.
Galeano, E. (1991) *The Book of Embraces*, New York: Norton & Co.
Galtung, J. (1976) 'The future: a forgotten dimension', in H. Ornauer, H. Wiberg, A. Sicinski and J. Galtung (eds) *Images of the World in the Year 2000*, Atlantic Highlands, NJ: Humanities Press.
Giddens, A. (1990) *The Consequences of Modernity*, Cambridge: Polity Press.
—— (1991) *Modernity and Self-identity*, Cambridge: Polity Press.
Gidley, J. (1998) 'Prospective youth visions through imaginative education', *Futures*, 30: 395–408.
Gillespie, K. and Allport, G. (1955) *Youth's Outlook on the Future: A Cross-national Study*, New York: Doubleday.
Gough, N. (1990) 'Futures in Australian education: tacit, token and taken for granted', *Futures*, 22: 298–310.
Griffin, D. (1988) *The Reenchantment of Science*, Albany, NY: State University of New York Press.

Griffin, S. (1994) *A Chorus of Stones: The Private Life of War*, London: The Women's Press.
*Guardian* (1992) 'Earth's future in balance say top scientists', *Guardian*, 27 February.
Hanna, N. (1998) *The Millennium: A Rough Guide to the Year 2000*, London: Rough Guides.
Harvey, D. (1989) *The Condition of Postmodernity*, Oxford: Blackwell.
Henderson, H. (1993) *Paradigms in Progress: Life beyond Economics*, London: Adamantine Press.
Henley Centre (1991) *Young Eyes: Children's Vision of the Future Environment*, London: Henley Centre for Forecasting.
Heron, J. (1989) *The Facilitators' Handbook*, London: Kogan Page.
—— (1990) *Helping the Client*, London: Sage Publications.
Hicks, D. (1976) 'Studying a world of change', *World Studies Bulletin*, 40: 9–15.
—— (1978) 'Perspectives on poverty and injustice: a course for student teachers', *The New Era*, 59: 58–62.
—— (1980) *Images of the World: An Introduction to Bias in Teaching Materials*, Occasional Paper No. 2, London: Centre for Multicultural Education, University of London Institute of Education.
—— (1981) *Minorities: A Teacher's Resource Book for the Multi-ethnic Curriculum*, London: Heinemann.
—— (ed.) (1988) *Education for Peace: Issues, Principles and Practice in the Classroom*, London: Routledge.
—— (1990) 'World Studies 8–13: a short history', *Westminster Studies in Education*, 13: 61–80.
—— (1994a) *Education for the Future: A Practical Classroom Guide*, Godalming: World Wide Fund for Nature UK.
—— (ed.) (1994b) *Preparing for the Future: Notes and Queries for Concerned Educators*, London: Adamantine Press.
—— (2001) *Citizenship for the Future: A Practical Classroom Guide*, Godalming: World Wide Fund for Nature UK.
Hicks, D. and Bord, A. (1994) 'Visions of the future: student responses to ecological living', *Westminster Studies in Education*, 17: 45–55.
Hicks, D. and Holden, C. (1995) *Visions of the Future: Why We Need to Teach for Tomorrow*, Stoke-on-Trent: Trentham Books.
Hicks, D. and Slaughter, R. (eds) (1998) *Futures Education: The World Yearbook of Education 1998*, London: Kogan Page.
Hicks, D. and Steiner, M. (eds) (1989) *Making Global Connections: A World Studies Workbook*, Edinburgh: Oliver & Boyd.
Hillcoat, J. (1996) 'Action research', chap. 12 in M. Williams (ed.) *Understanding Geographical and Environmental Education: The Role of Research*, London: Cassell.
Hitchcock, G. and Hughes, D. (1989) *Research and the Teacher*, London: Routledge.
Hobsbawm, E. (1994) *Age of Extremes: The Short Twentieth Century 1914–1991*, London: Michael Joseph.
Holden, C. (1989) 'Teaching about the future with younger children', in R. Slaughter (ed.) *Studying the Future: An Introductory Reader*, Canberra: Commission for the Future and Bicentennial Authority.
Huber, B. (1978) 'Images of the future', in J. Fowles (ed.) *Handbook of Futures Research*, Wesport, CT: Greenwood Press.

Huckle, J. (1990) 'Environmental education: teaching for a sustainable future', chap. 10 in B. Dufour (ed.) *The New Social Curriculum*, Cambridge: Cambridge University Press.
Huckle, J. and Sterling, S. (eds) (1996) *Education for Sustainability*, London: Earthscan.
Hutchinson, F. (1996) *Educating beyond Violent Futures*, London: Routledege.
Inayatullah, S. (1993) 'From 'who am I?' to 'where am I?' Framing the shape and time of the future', *Futures*, 25: 235–53.
—— (ed.) (1996) 'What futurists think', special issue of *Futures*, 28 (6/7).
Inglehart, R. (1997) *Modernization and Postmodernization: Cultural, Economic and Political Change in 43 Societies*, Princeton, NJ: Princeton University Press.
Jay, M. (1993) *Force Fields: Between Intellectual History and Cultural Critique*, London: Routledge, p. 84.
Johnson, L. (1987) 'Children's visions of the future', *The Futurist*, 21: 36–40.
Jones, C. (1998) 'Planet eaters or star makers? One view of futures studies in higher education', *American Behavioral Scientist*, 42: 470–83.
Jouvenel, B. de (1964) *The Art of Conjecture*, New York: Basic Books.
Jungk, R. and Galtung, J. (1969) *Mankind 2000*, London: Allen & Unwin.
Jungk, R. and Mullert, N. (1987) *Future Workshops: How to Create Desirable Futures*, London: Institute for Social Inventions.
Kahn, H. and Wiener, A. (1967) *The Year 2000*, New York: Macmillan.
Kaza, S. (1999) 'Liberation and compassion in environmental studies', in G.A. Smith and D.Williams (eds) *Ecological Education in Action: On Weaving Education, Culture and the Environment*, Albany, NY: State University of New York Press.
Kellert, S.R. and Wilson, E.O. (eds) (1993) *The Biophilia Hypothesis*, Washington, DC: Island Press/Shearwater.
Kermode, F. (1968) *The Sense of an Ending*, London: Oxford University Press.
—— (1995) 'Waiting for the end', chap. 11 in M. Bull (ed.) *Apocalypse Theory and the Ends of the World*, Oxford: Blackwell.
Kumar, K. (1991) *Utopianism*, Milton Keynes: Open University Press.
—— (1995a) *From Post-industrial to Post-modern Society*, Oxford: Blackwell.
—— (1995b) 'Apocalypse, millennium and utopia today', chap. 9 in M. Bull, (ed.) *Apocalypse Theory and the Ends of the World*, Oxford: Blackwell.
Kurian, G. and Molitor, G. (eds) (1996) *Encyclopedia of the Future*, vols 1 and 2, New York: Simon & Schuster.
Landes, R. (1997) 'The apocalyptic year 1000', chap. 2 in C. Strozier and M. Flynn, *The Year 2000: Essays on the End*, New York: New York University Press.
Lather, P. (1986) 'Research as praxis', *Harvard Educational Review*, 56: 257–77.
—— (1991) *Getting Smart: Feminist Research and Pedagogy with/in the Postmodern*, London: Routledge.
Le Guin, U. (1988) *Always Coming Home*, London: Grafton Books.
Lifton, R.J. (1992) 'From a genocidal mentality to a species mentality', in S. Staub and P. Green (eds) *Psychology and Social Responsibility: Facing Global Challenges*, New York: New York University Press.
Lincoln, Y.S. and Denzin, N.K. (2000) 'The seventh moment: out of the past', chap. 41 in N.K. Denzin and Y.S. Lincoln (eds) *Handbook of Qualitative Research*, 2nd edn, Thousand Oaks, CA: Sage Publications.
Lister, I. (1987) 'Global and international approaches in political education', in C. Harber (ed.) *Political Education in Britain*, Lewes: Falmer Press.

Littledyke, M. (1996) 'Science education for environmental awareness in a postmodern world', *Environmental Education Research*, 2: 197–214.
Livingstone, D. (1976) 'Images of the educational future in advanced industrial society: an Ontario enquiry', *Canadian Journal of Education*, 1: 13–29.
—— (1983) 'Intellectual and popular images of the educational and social future', *Class Ideologies and Educational Futures*, London: Falmer Press.
Longstreet, W. and Shane, H. (1993) *Curriculum for a New Millennium*, Boston, MA: Allyn & Bacon.
Lowenthal, D. (1995) 'The forfeit of the future', *Futures*, 27: 385–95.
McGrew, A. (1992) 'A global society?' chap. 2 in S. Hall, D. Held, and T. McGrew (eds) *Modernity and Its Futures*, Cambridge: Polity Press.
McKibben, B. (1995) *Hope, Human and Wild: True Stories of Living Lightly on the Earth*, Boston, MA: Little, Brown & Co.
McLaughlin, C. and Davidson, C. (1985) *Builders of the Dawn: Community Lifestyles in a Changing World*, Summertown, TEN: Book Publishing Co.
Macy, J. and Brown, M. (1998) *Coming Back to Life: Practices to Reconnect Our Lives, Our World*, San Gabriola Island, BC: New Society Publishers.
Markley, O. (1992) 'Using in-depth intuition in creative problem solving and strategic innovation', in S. Parnes (ed.) *Source Book for Creative Problem-solving*, Buffalo, NY: Creative Education Foundation Press.
Masini, E. (1987) 'Women as builders of the future', *Futures*, August: 431–6.
Masser, I., Sviden, O. and Wegener, M. (1992) *The Geography of Europe's Futures*, London: Belhaven Press.
May, G. (1996) *The Future is Ours: Foreseeing, Managing and Creating the Future*, London: Adamantine Press.
Meadows, D.H., Meadows, D.L. and Randers, J. (1972) *The Limits to Growth*, New York: Universe.
Meadows, D., Meadows, D. and Randers, J. (1992) *Beyond the Limits: Global Collapse or a Sustainable Future*, London: Earthscan.
Merchant, C. (1992) *Radical Ecology*, London: Routledge.
Metcalf, B. (1996) *Shared Visions, Shared Lives: Communal Living around the Globe*, Forres: Findhorn Press.
Milbrath, L. (1989) *Envisioning a Sustainable Society: Learning Our Way Out*, Albany, NY: State University of New York Press.
Moll, P. (1991) *From Scarcity to Sustainability: Futures Studies and the Environment: The Role of the Club of Rome*, Frankfurt: Peter Lang.
Moltmann, J. (1967) *The Theology of Hope*, London: SCM Press.
—— (1975) *The Experiment of Hope*, London: SCM Press.
Nanus, B. (1992) *Visionary Leadership*, San Francisco: Jossey-Bass.
O'Rourke, B. (1994) 'Futures and the curriculum', in *Shaping the Future: Review of the Queensland School Curriculum*, vol. 3, Brisbane, State of Queensland.
Ornauer, H., Wiberg, H., Sicinski, A. and Galtung, J. (eds) (1976) *Images of the World in the Year 2000*, Atlantic Highlands, NJ: Humanities Press.
Orr, D. (1992) *Ecological Literacy: Education and the Transition to a Postmodern World*, Albany, NY: State University of New York.
Page, J. (2000) *Reframing the Early Childhood Curriculum: Educational Imperatives for the Future*, London: Routledge Falmer.
Piercy, M. (1976) *Woman on the Edge of Time*, London: The Women's Press.

Pike, G. and Selby, D. (1988) *Global Teacher, Global Learner*, London: Hodder & Stoughton.
Polak, F. (1955) *The Image of the Future*, vols 1 and 2, New York: Oceana.
—— (1972) *The Image of the Future*; trans. and abr. by E. Boulding, San Francisco: Jossey-Bass/Elsevier.
Popkin, R. (1995) 'Seventeenth-century millenarianism', chap. 6 in M. Bull (ed.) *Apocalypse Theory and the Ends of the World*, Oxford: Blackwell.
Postel, S. (1992) 'Denial in the decisive decade', chap. 1 in L. Brown and L. Starke (eds) *State of the World 1992*, London: Earthscan.
Progoff, I. (1975) *At a Journal Workshop*, New York: Dialogue House.
Rawling, E. (1992) *Programmes of Study: Try This Approach*, Sheffield: Geographical Association.
Real World Coalition (1996) *Politics of the Real World*, London: Earthscan.
—— (2001) *From Here to Sustainability: Politics in the Real World*, London: Earthscan.
Reardon, B. (1985) *Sexism and the War System*, New York: Teachers College Press.
Redclift, M. and Benton, T. (eds) (1994) *Social Theory and the Global Environment*, London: Routledge.
Reinharz, S. (1992) *Feminist Methods in Social Research*, New York: Oxford University Press.
Richardson, R. (1990) *Daring to be a Teacher: Essays, Stories and Memoranda*, Stoke-on-Trent: Trentham Books.
—— (1996) *Fortunes and Fables: Education for Hope in Troubled Times*, Stoke-on-Trent: Trentham Books.
Rickinson, M. (2001) 'Learners and learning in environmental education: a critical review', *Environmental Education Research*, 7: 207–317.
Riley, K. (1989) *Toward Tomorrow*, New York: Scholastic.
Rogers, M. (1994) *Learning about Global Futures: An Exploration of Learning Processes and Changes in Adults*, DEd thesis, Toronto: University of Toronto.
—— (1998) 'Student responses to learning about futures', chap. 15 in D. Hicks and R. Slaughter (eds) *Futures Education: The World Yearbook of Education 1998*, London: Kogan Page.
Rogers, M. and Tough, A. (1992) 'What happens when students face the future?' *Futures Research Quarterly*, 8: 9–18.
—— (1996) 'Facing the future is not for wimps', *Futures*, 28: 491–6.
Rosenau, P. (1992) *Post-modernism and the Social Sciences*, Princeton, NJ: Princeton University Press.
Roszak, T., Gomes, M. and Kanner, A. (eds) (1995) *Ecopsychology: Restoring the Earth, Healing the Mind*, San Francisco: Sierra Club.
Sargisson, L. (1996) *Contemporary Feminist Utopianism*, London: Routledge.
Schaffer, E. (1995) 'Secular apocalypse: prophets and apocalyptics at the end of the eighteenth century', chap. 7 in M. Bull (ed.) *Apocalypse Theory and the Ends of the World*, Oxford: Blackwell.
Schwartz, P., Leyden, L. and Hyatt, J. (1999) *The Long Boom*, Reading, MA: Perseus Books.
Seager, J. (1993) *Earth Follies*, London: Earthscan.
Sessions, G. (1995) *Deep Ecology for the 21st Century*, Boston, MA: Shambala.
Shiva, V. (1989) *Staying Alive: Women, Ecology and Development*, London: Zed Books.
Slaughter, R. (1985) *What Do We Do Now the Future Is Here?* Lancaster: University of Lancaster.
—— (1991) 'Changing images of futures in the 20th century', *Futures*, 23: 499–515.

—— (1995) *The Foresight Principle*, London: Adamantine Press.
—— (ed.) (1996) *The Knowledge Base of Futures Studies*, 3 vols, Melbourne: DDM Media Group.
—— (2000) *Futures for the Third Millennium: Enabling the Forward View*, St Leonards, NSW: Prospect Media Pty Ltd.
Smith, G.A. (1992) *Education and the Environment: Learning to Live with Limits*, Albany, NY: State University of New York Press.
Spradlin, A., Daniell, B., Hannegan, D. and Spangle, M. (1989) 'A quantitative assessment of the Imagining a World without Weapons Workshop', paper presented at the Annual Conference of Consortium for Peace Research and Education, Denver, CO, 7 October.
Starhawk (1990) *Dreaming the Dark: Magic, Sex and Politics*, London: Mandala/ Unwin.
—— (1993) *The Fifth Sacred Thing*, New York: Bantam Books.
Staub, S. and Green, P. (eds) (1992) *Psychology and Social Responsibility: Facing Global Challenges*, New York: New York University Press.
Steiner, M. (1993) *Learning from Experience: Cooperative Learning and Global Education*, Stoke-on-Trent: Trentham Books.
—— (ed.) (1996) *Developing the Global Teacher: Theory and Practice in Initial Teacher Education*, Stoke-on-Trent: Trentham Books.
Sterling, S. (2001) *Sustainable Education: Re-visioning Learning and Change*, Totnes: Green Books.
Stotland, E. (1969) *The Psychology of Hope*, San Francisco: Jossey-Bass.
Swimme, B. and Berry, T. (1994) *The Universe Story*, San Francisco: HarperSan Francisco.
Tedlock, B. (2000) 'Ethnography and ethnographic representation', chap. 17 in N.K. Denzin and Y.S. Lincoln (eds) *Handbook of Qualitative Research*, 2nd edn, Thousand Oaks, CA: Sage Publications.
Thick, C. (ed.) (1996) *The Right to Hope: Global Problems, Global Visions*, London: Earthscan.
Thompson, D. (1996) *The End of Time: Faith and Fear in the Shadow of the Millennium*, London: Sinclair-Stevenson.
Toffler, A. (1970) *Future Shock*, New York: Random House.
—— (1974) *Learning for Tomorrow: The Role of the Future in Education*, New York: Vintage Books/Random House.
Trainer, T. (1995) *The Conserver Society*, London: Zed Press.
Trimby, P. (1995) *Solar Water Heating: A DIY Guide*, Machynlleth: Centre for Alternative Technology.
Vaughn, S., Schumm, J. and Sinagub, J. (1996) *Focus Group Interviews in Education and Psychology*, London: Sage Publications.
Vilgot, O. (1995) 'Pupils' views of the future', in A. Osler, H-F. Rathenow and H. Starkey (eds) *Teaching for Citizenship in Europe*, Stoke-on-Trent: Trentham Books.
Vision 21 (1997) *Sustainable Gloucestershire*, Cheltenham: Vision 21.
Wagar, W. (1992) *The Next Three Futures: Paradigms of Things to Come*, London: Adamantine Press.
Walford, R. (1984) 'Geography and the future', *Geography*, 69: 193–208.
—— (2001) 'Radical responses, 1975–85', chap. 9 in *Geography in British Schools 1850–2000*, London: Woburn Press.
Walker, J.T. (1996) 'Postmodernism and the study of the future', *Futures Research Quarterly*, 12: 51–70.

Walsh, R. (1992) 'Psychology and human survival: psychological approaches to contemporary global threats', in S. Staub and P. Green (eds) *Psychology and Social Responsibility: Facing Global Challenges*, New York: New York University Press.
Wark, K. (1996) 'Tech noir cinema and the "techno-fear" future', in R. Slaughter (ed.) *The Knowledge Base of Futures Studies*, Melbourne: DDM Media Group.
Weedon, C. and Light, J. (1995) *The Reed Beds at C.A.T.*, Machynlleth: Centre for Alternative Technology.
Weisbord, M. and Janoff, S. (1995) *Future Search: An Action Guide to Finding Common Ground in Organizations and Communities*, San Francisco: Berrett-Koehler.
Worcester, R. (1994) *The Sustainable Society: What We Know about What People Think and Do*, London: MORI.
Wright, D. (1998) 'Studying the near future in geography', *Teaching Geography*, 23: 42.
Yorkshire Dales National Park Committee (1989) 'Landscapes for Tomorrow', Skipton: Yorkshire Dales National Park.
Yothu Yindi (1992) 'Treaty', a track from *Tribal Voice*, Mushroom Records.
Ziegler, W. (1987) *Designing and Facilitating Projects and Workshops in Futures-Invention*, Boulder, CO: Futures Invention Associates.
—— (1989) *Envisioning the Future: A Mindbook of Exercises for Futures-Inventors*, Boulder, CO: Futures-Invention Associates.
—— (1991) 'Envisioning the future', *Futures*, 23: 516–27.

# Index

action dimension 101
action: finding the path of 102
action readiness 67
activist academics 99
affective dimension 101
affirmation 13
age variations 32–6
Agricultural Revolution 68
alienation 99
alternative energy 65, 66
alternative futures 7–8, 15, 17, 46, 54
alternative technology *see* appropriate technology
anarchism 4, 5, 6, 9
anticipating change 16
anthropology 114
anti-nuclear movement 9
anti-utopia 28
apocalypse 110, 111, 121
appropriate technology 91, 92, 93, 96, 118
archaeology of the future 78
atmosphere 118
Auschwitz 69
Australian Science, Technology and Engineering Council 38
autobiographical writing *see* journalling
autobiography 1, 2, 72, 75, 124
awakening: the heart 102; the mind 102; the soul 102
awe and wonder 73

Ballard, J.G. 60, 67
Bardwell, L. 69
baseline future 58, 63, 66, 67, 85, 120
Beare, H. 54
Beck, U. 99
Behar, R. 113
Bell, D. xiv, xv
Bell, W. xi–xvi, 17, 54, 60, 79, 123

Berry, T. 68
better world 4, 17, 56, 67
biophylia hypothesis 77, 94
Bloch, E. 70, 80
Bord, A. xvii
Borough Road College 4
Boulding, E. xvii, 58, 60, 62, 64, 67, 79, 82, 85, 118, 120
breach in time 62
Brown, L. 69
Brown, M. 29, 31

Callenbach, E. 88
Campaign for Nuclear Disarmament 3, 6
campaigning 4
Cantril, H. 27
catalytic validity 75
Centre for Alternative Technology 52, 90–7; website 91
Centre for Global Education 129, 176
Centre for Peace Studies 6
changing times 123
children's concerns *see* children's views
children's views 20–1, 29–39
Charlotte Mason College 5
Christian Aid 127
citizenship 19, 40
citizenship education 8, 18
clarifying values 16
Clarke, I.F. 61
Club of Rome xiv, 12
cognitive dimension 100, 108
collective struggles 75, 84
community 8, 9, 23, 33, 38, 39, 43, 65, 66, 82, 83, 96, 121; lifestyles 70, 92, 95
communities of hope 85
concentration camps 3
conscious democracy 53
cooperative inquiry 114

conserver society 87
construct validity 75
conviviality 65, 83, 85
Cornish, E. 127
Cottle, T. 78
Council for Education in World Citizenship 126
Creating Preferable Futures 129
creative imagination 16, 17, 64
creativity *see* human creativity
critical reflection 13
critical thinking 16, 17

Dan Dare xvii
Danziger, K. 27
data triangulation 71
Dator, J. 54, 58, 61, 79, 122
decision making 16, 17
degradation of planet 72
demilitarisation 31
denial 70, 79, 99, 100, 105
Denzin, N. 80, 81, 113, 121, 123
desirable futures *see* preferable futures
despair 22, 62, 63, 72, 112
development education 9, 13, 40, 126, 129
Development Education Association 127
dimensions of learning 100, 106
disengaged observer 2
Dobson, S. 87
dominant social paradigm 9, 87
Dovey estuary 90
Dulas valley 90, 91
dystopia 2, 76, 88, 111

Earth Summit 1992 11; 2002 32
Eastern philosophy 1
Eckersley, R. 31, 38
eco-cabins 91, 92, 95
eco-feminism 7
ecopsychology 99
ecotopia 120
education 65; process of 13
educational rationale 16–17, 123
education for the future *see* futures education
Elgin, D. 53, 97
emancipatory knowledge 113; movements 87
empowerment 70, 107, 108
empowerment dimension 101, 107, 108
end of time 110
energy conversions 93
Enlightenment 69, 108, 111

environmental action 4; attitudes 100; education 13, 18, 40, 54, 59, 69, 74, 77, 98, 99, 107, 129
envisioning 15, 19, 53–9, 61–3, 82, 85, 118, 121, 123
equality 82
equity 60, 65
ethnography 2, 81, 82, 113, 102, 124
evaluation 66
Everitt, A. 28, 111
everyday life 14
existential anxieties 120; dilemmas 9, 99; dimension 100, 107, 108
experiments from the future 9
extended present 44–5

face validity 75
Facilitator Styles 8
faith and belief 75, 84
family stories 116, 119, 121
feminist perspectives 19
feminist researchers 2
Ferns, D. 8
fieldwork 90–7
Fien, J. 6
*fin de siecle* 111, 112
focus group 72, 74, 81, 82, 113, 119
food 65, 118
forecasting xiii, xiv
formative influences 1
Fox, M. 8, 9
Freire, P. 5, 76, 121
future consequences xiii; generations 15, 17, 39, 100; geographies 42–7; social 28; technological 28
futureless education 66
futures dimension 2, 13, 14–20, 38, 52, 122, 123, 127–30; rationale for 15
futures education 13, 17, 38, 40, 60, 68, 70, 123, 124, 128, 129, 130; field 8, 54; movements 54; research 54, 99; thinking xiii, xiv, xv, xvi, 18, 28, 61, 64; workshops 39, 54, 56–9, 62, 81
Futures Project 8
futures studies xiv, 41, 54, 55, 61, 78, 79, 98, 108, 123, 130; departments of xvi
futurists xiv, xv, 26, 27, 54, 56, 60, 62, 79, 86

Gabor, D. xiv
Galeano, E. 23
Galtung, J. xiv, 5, 28, 55, 56, 79
gender 36–7, 82

geography 1, 3, 4, 5, 9, 18, 40–52, 127; of the future 40
Giddens, A. 69, 98
Gidley, J. 38
global dimension 5, 37, 38, 12, 125, 126–7; education 2, 7, 9, 13, 68, 69, 71, 74, 75, 98, 107, 126, 129; futures 90, 98–108; issues 71, 98, 100, 104, 107, 108, 123;
global educators *see* global education
global perspective *see* global dimension
Global Teacher Project 7
globalisation 69, 90, 99, 112, 117
Gloucestershire 2, 4
Gough, N. 15, 129
grammar school 1, 3
green 65, 85
Greenpeace 2, 12
Griffin, D. 76, 87, 121
Guildford 8

Hammond, J. xvii
Harvey, D. 69
Hawaii Research Centre for Futures Studies 55
Hawkwood xvii
health 65
Henderson, H. xvii, 54, 79
Henderson, J. 126
Henley Centre for Forecasting 30
high consequence risk 69
History: departments of xvi
Hobsbawm, E. 67, 110
Holden, C. xvii, 29
holistic learning 98, 108; paradigm 9
holocaust 56
hope 12, 67, 70, 71, 76, 77, 79, 80, 83, 84, 85, 88, 112, and fears 27, 30, 31, 32–35, 38, 41, 42, 61, 66, 68
Huber, B. 60
Huckle, J. 69, 76
human: creativity 75, 84, 119; nature 73; rights 72, 82
humour 3, 75, 84
hunger and poverty 72
Hutchinson, F. xvii, 6, 30, 38

image and action 124
images of the future xii, 14, 15, 19, 21–25, 26, 27–28, 37, 38, 55, 58, 60, 79, 118, 123, 124
imaging *see* envisioning
imaging capacity 22

Inayatullah, S. xvii, 78
incomprehension 122
indoctrination 6
Industrial Revolution 23, 69
Inglehart, R. 87
initial teacher education 5
injustice: sense of 3, 9
interdependence 126
intergenerational equity 31; justice 19
International Peace Research Association 5
International Year of Peace 6

Johnson, L. 30, 31
Jones, C. 108
journalling 71, 72, 81, 86, 96, 102, 107, 114
journeys 1, 3, 8, 105, 124
Jouvenal, B. de xiv
Judge Dredd 28
Jungk, R. xiv, 56, 57, 81
justice 65, 66

Kaza, S. 108
Kermode, F. 112
Klineberg, S. 78
Kumar, K. 23, 67, 88

Lancaster 5
Lather, P. 113
Leeds Metropolitan University 7
Le Guin, U. 78, 80
life 119
Lifton, J. 70
Lincoln, Y. 80, 113
Lister, I. 13
Livingstone, D. 28, 60
Local Agenda 21, 52, 87, 91
local futures 33–4
local-global connections 126
Lowenthal, D. 61

Machynlleth 91
McKibben, B. 71, 76
Macy, J. 7, 8, 9, 99, 108
Masini, E. 24, 79
May, G. 54
Meadows, D. xv, 58, 61, 97
meaning of life 3
media influences 124
memorabilia 115
mentors and colleagues 75, 85
Merchant, C. 69

metanarrative 69, 87
mid-life crisis 104
Milbrath, L. 87
millenarianism 110, 112, 121
millennial cults/movements *see* millenariansim
millennial tradition 109, 121
millennium 12, 28, 60, 67, 88, 109–21
Milojevic, I. xvii
Minority Rights Group 5
modernity 77, 87, 111, 113
Moll, P. 28
Moltmann, J. 70, 71, 80
moral analysis xiii
Morris, W. 9, 22, 88, 130
multicultural education 9, 18
multinational corporations 72

National Academy of Sciences 11
natural world 75, 84, 119
Network for Nuclear Concern 6
new environmental paradigm 87
new movements in education 13
new social movements 8
non-violent direct action 8
Nova Scotia 2
nuclear accidents 72
nuclear arms race 6

Only One earth 5
Ontario Institute for Studies in Education 100
ontological need 76
optimism 36, 37, 38, 55, 69, 70, 71, 99, 105, 106, 108
organic horticulture 65, 91, 92, 94
Ornauer, H. 31, 55
O'Rourke, B. 38
Orr, D. 39
other people's lives 75, 84
Oxfam 127

Page, J. xvii
paradigm shift 8, 68, 87
paradox: of the future 127; of the past 128; of the present 128
participatory exploration 114
peace 31, 65; education 5, 6, 7, 9, 13, 129
peaceful future 65, 85
people 118, 119
perennial questions 120
personal: futures 31–3, 36; growth 9; insecurity 99; narrative 1, 9

person-centred education 5, 86
pessimism 28, 36, 37, 38, 55, 62, 64, 69, 99, 100, 104, 106
Piercy, M. 88
Polak, F. xii, xiv, 22, 38, 55, 56, 58, 61, 79
political Right 6
politicisation 4, 9
popular movies 61
possible futures xiii, xvi
positivist paradigm 113
Postel, S. 24, 70
post-materialist society 87
postmodern education 39
postmodernism 87; deconstructive 75, 76, 86; revisionary 75, 76, 86, 88
postmodernity 69, 81, 86, 98, 117
practical application 123
preferable futures xiii, xiv, 15, 18, 21, 31, 41, 53–54, 55, 61, 62, 64, 79, 82, 91, 118, 120, 121
primary school 2, 36
probable futures xiii, xiv, 15, 18, 21, 41, 46, 64
prospective thinking xiv
psychic numbing 70, 99
psychology of despair 68, 77
pull of the future 21
pupil motivation 16
push of the past 21

qualitative research 80, 113, 124
quality of life 82
question of hope 70
questions about the future 20, 21

racist bias 5
radical education 8
radicalism 70
Real World Coalition 12
Reardon, B. 6
reed-bed sewage system 91, 94
Reinharz, S. 71
relationships 75, 84
repression 99
research: needed 124; process 86
Resnais, A. 3
resistance 9
resources and policy 124
responsible citizens 13, 17
Richardson, R. xvii, 5, 7, 13–14, 22, 77, 126
Rickinson, M. 59
Rogers, C. 5

Rogers, M. xvii, 38, 62, 79, 100, 102, 106, 107
Roszak, T. 99
roots 84, 85
Royal Society 11

St Martins College of Higher Education 6
Sargisson, L. 88
scenarios 47–52, 57, 101
Schwartz, P. xv
Scientific Revolution 68, 69
self-discovery 3
sense of self 75, 84, 101
sexuality 8
secondary school 36
Shiva, V. 23
Slaughter, R. xv, xvii, 17, 21, 54
Smoker, P. 5
social education 58, 107
socially committed education *see* socially critical education
socially critical education 1, 5, 25, 76, 81, 108
solar water heaters 94
soul-searching 101
sources of hope 70, 71–5, 77, 81, 84, 85, 88, 108, 118, 123
spatial dimension 8, 40, 90, 126, 127
spirit of place 95
spirituality 7, 8, 9, 38
Starhawk 88
Steiner education 38
Stevenson, T. xvii
stories of hope 62, 68–77
students' preferable futures 65
success stories 76, 101, 108
sustainable living 91
sustainable futures 8, 9, 12, 15, 19, 24, 52, 52, 53, 58, 62, 68, 86, 97, 88,
sustainable society *see* sustainable futures
synergy 81

taken-for-granted futures 17
tacit futures 17
teacher education 6–7, 9
technocratic dreaming 31
temporal dimension 8, 40, 90, 127
therapy 7, 8, 9, 70
Third World women 23
Thompson, D. 109
threats to survival xi, xvi
three awakenings 102
time 119
timelines 45–7, 64
Toffler, A. xiv, 12, 14, 29, 128
Toh, S-H. 6
token futures 17
Tough, A. 62, 79
townscapes 82, 83
Trainer, T. 87
transportation 65, 82, 85, 118
triangulation 75
trends 12, 64

UK survey 31–7
US Council for Social Studies 128
University of Exeter 3
University of London Institute of Education 8
University of Surrey 8
utopia 9, 22, 56, 71, 88, 120, 121
utopian communities 23
utopian studies 70
utopian tradition 18, 19, 86, 87, 88
utopians 13

validation 75
views of the future *see* images of the future
Villanueva, C. xvii
Vision 21, 87
visionaries 75, 85
visions of the future xiii, 22, 39, 52, 54, 92
visualisation 54, 57, 63, 64, 65, 66, 67, 82
vulnerable observer 120

Wagar, W. 60, 124
Walker, J.T. 87
Walsh, R. 99
Ward, R. xvii
Wark, K. 61
water consumption 91
Western cosmology 110
Whitaker, P. xvii
wholeness: in individuals 13; in society 13
Wilson, E. 94
Windscale 5
wish list 62, 82
witnesses from the future 80
wood consumption 91
Wood, K. xvii
women: as builders of the future 24
women's movement 5, 9
work 65
Worcester, R. 87
World Future Society xv, 54, 128

World Futures Studies Federation xv, 54, 56, 128
World Images 2000 Project 28
world studies 7, 8, 9, 126; see *also* global education
World Studies 8–13, 6, 7, 126, 128
World Studies Project 5, 176
Worldwatch Institute 12

World War Two 3, 116
World Wide Fund for Nature UK xvii, 8
Yothu Yindi 23
young people: concerns 123; fears 31; images 27, 29, 37

Ziegler, W. 57, 58, 62